LABORATORY MANUAL

Conceptual Physical Science

SIXTH EDITION

Paul G. Hewitt
John Suchocki
Leslie A. Hewitt
Dean Baird

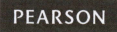

Boston Columbus Indianapolis New York San Francisco Upper Saddle River
Amsterdam Cape Town Dubai London Madrid Milan Munich Paris Montréal Toronto
Delhi Mexico City São Paulo Sydney Hong Kong Seoul Singapore Taipei Tokyo

Editor-in-Chief: Jeanne Zalesky
Executive Editor: Scott Dustan
Director of Marketing: Christy Lesko
Program Manager: Mary Ripley
Program and Project Management Team Lead: Kristen Flathman
Production Service: Cenveo Publisher Services, Rose Kernan
Compositor: Cenveo Publisher Services
Manufacturing Buyer: Maura Zaldivar-Garcia
Printer and Binder: LSC Communications

ISBN-10-digit: 0-13-409141-8
ISBN-13-digit: 978-0-13-4091419

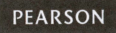

Contents

Introduction

Many students enter this course with very few hands-on science learning experiences. The activities and experiments in this manual are designed to provide those experiences. Whether it's attempting to use a battery and wires to get a bulb to light, collecting carbon dioxide by water displacement, or typing minerals on the basis of observable characteristics, these explorations engage students in direct interaction with the natural world in ways that will deepen both their curiosity and their understanding.

It has been suggested that we retain about 5% of what we read and about 75% of what we do. Clearly, the activities and experiments in this manual play an important role in the complete *Conceptual Physical Science* program of instruction. Enjoy!

Acknowledgments

Some of the physics experiments and activities in this manual first appeared in the *Conceptual Physics* lab manuals authored more than a decade ago by Paul Robinson of San Mateo High School (retired) in San Mateo, California.

Pushing Things Around, Worlds of Wonder, Bouncing off the Walls, Pool Cubes, Faraday's Electromagnetic Lab, Water Waves in an Electric Sink, Pixel Peeping, and *Radioactive Speed Dating* involve the use of computer simulations developed and distributed by the PhET team at the University of Colorado, Boulder. Their simulations are robust, engaging, accurate, and free. Thank you, PhET!

For *Walking the Plank,* we are grateful to Howie Brand, formally of Suitland High School in Forestville, Maryland.

Thanks to Ted Brattstrom and Mary Graff for *Mystery Powders.* For *Salt and Sand,* we are grateful to Erwin W. Richter.

For contributions to nearly all the earth science labs, we are grateful to Leslie's husband, Bob Abrams (also a geologist). For valuable feedback and suggestions, we thank the late City College of San Francisco physics and geology instructor Jim Court. For other geology resources, we are grateful to the American Geological Institute and the National Association of Geology Teachers.

For astronomy activities and experiments, we thank Ted Brattstrom for *Reckoning Latitude,* and Forrest Luke for *Tracking Mars.* Both Ted and Forrest were high school teachers and amateur astronomers in Hawaii—Ted at Pearl City High School, and Forrest at Leilehua High School. For valuable feedback and advice on most of the astronomy material, we are indebted to the late Richard Crowe, University of Hawaii at Hilo, and to John Hubisz, North Carolina State University.

For general suggestions on all aspects of this manual, we remain grateful to Charlie Spiegel and Marshall Ellenstein.

Important Notes About Safety

When performing the activities and experiments in this manual, you should always keep your own safety and that of others in mind. Most safety rules that must be followed involve common sense. For example, if you are ever unsure of what to do or the details of the lab materials, ask your instructor, who should always be there to help you.

To guide you to safe practices, in this manual you will find certain icons posted by selected activities. Here are those icons and what they indicate:

 Wear approved safety goggles. Wear goggles when working with a chemical or solution, or when heating substances.

 Wear gloves. Wear gloves when working with chemicals.

 Flame/heat. Keep combustible items, such as paper towels, away from any open flame. Handle hot items with tongs, oven mitts, potholders, or the like. Do not put your hands or face over any boiling liquid. Use only heat-proof glass, and never point a heated test tube or other container at anyone. Turn off the heat source when you are finished with it.

Here are some specific safety rules that you should follow at all times:

1. Do not eat or drink in the laboratory.

2. Maintain a clean and orderly work space. Clean up spills at once, or ask for assistance in doing so.

3. Do not perform unauthorized experiments. First obtain permission from your teacher. It is most important that others know what you are doing and when you are doing it. Think of it this way: If you are doing an experiment alone and something goes terribly wrong, there will be no one there to help you. That's not using your common sense.

4. Do not taste any chemicals or directly breathe any chemical vapors.

5. Check all chemical labels for both name and concentration.

6. Do not grasp recently heated glassware, clamps, or other heated equipment, because they remain hot for quite a while.

7. Discard all excess reagents or products in the proper waste containers.

8. If your skin comes in contact with a chemical, rinse under cold water for at least 15 minutes.

9. Do not work with flammable solvents near an open flame.

10. If you are not certain whether or not a chemical is hazardous, assume that it is.

Master List of Apparatus

Item	Activities and Experiments
acetate strips	Electroscopia
acetic acid	Smells Great!
acetone	Chemical Personalities
acrylic tube	Dropping the Ball
air core solenoid	Generator Activator
alcohol, rubbing	Mystery Powders, Circular Rainbows
aluminum foil	I'm Melting! I'm Melting!, The Lemon Electric (optional), Solar Power I
ammeter, DC	Ohm, Ohm on the Range
ammonia cleanser	Sensing pH
baking soda	Mystery Powders, Bubble Round-Up, Sensing pH, Why Winds Wind Around Earth
balance (precision)	Blowout (optional), The Fountain of Fizz, Upset Stomach
balance, equal arm	Spiked Water
ball, lead	The Newtonian Shot
ball, steel	Dropping the Ball, Bull's Eye
balloons	Charging Ahead
barium chloride powder	Bright Lights
batteries, small (AAAA- or N-cell)	The Lemon Electric
batteries, 9-volt	Sensing pH
batteries, D-cell	Batteries and Bulbs, Motor Madness, The Lemon Electric
battery, ignition (No. 6, 1.5-V dry cell)	The Lemon Electric
battery, lantern (6-volt)	Electric Magnetism
BB shot	Thickness of a BB Pancake
beakers	Mystery Powders, Circular Rainbows
beakers, 1000-mL	Bubble Round-Up
beakers, 250-mL	Chemical Personalities, Tubular Rust
beakers, 500-mL	Chemical Personalities, Soil and Water
benzyl alcohol	Smells Great!
boiling chips	Chemical Personalities
borax	Sensing pH
boric acid	Mystery Powders
bottle (dropping)	Hot Spot Volcanoes
bottles, glass ketchup	Pure Sweetness
Bouncing Dart	Things That Go Bump
box (food storage)	Hot Spot Volcanoes
brown sugar	Pure Sweetness
bubble solution	Charging Ahead
bucket, 3-gallon	Temperature Mix, Stream Study
bulb sockets, miniature	Ohm, Ohm on the Range, Batteries and Bulbs
bulbs, miniature	Ohm, Ohm on the Range, Batteries and Bulbs
Bunsen burner	Bright Lights
buret clamps	Dropping the Ball, Upset Stomach
buret stands	Upset Stomach

calcium chloride powder	Bright Lights
calcium chloride solution	Chemical Personalities
can, empty	Bull's Eye
cans, radiation	Canned Heat: Heating Up, Canned Heat: Cooling Down
cards, 3" x 5"	Sunballs
clamps, right angle	Pole-Arizer
club soda	Sensing pH
collar hooks	Walking the Plank, The Weight, Motor Madness
colored pencils	Bright Lights
compasses, geometric	Over and Under, Tracking Mars
compasses, navigational	Electric Magnetism
computer	Sonic Ranger, Pushing Things Around, Putting the Force Before the Cart, Worlds of Wonder, Pool Cubes: Density, Pool Cubes: Buoyancy, Bouncing Off the Walls, Faraday's Electromagnetic Lab, Water Waves in an Electric Sink, Pixel Peeping
cooking pot	Pure Sweetness, Sensing pH
copper acetate	Crystal Growth
cornstarch	Mystery Powders
crossbars (short rods)	Walking the Plank, The Weight, Pole-Arizer
crucible	Circular Rainbows
cupric chloride powder	Bright Lights
cupric sulfate pentahydrate crystals	Chemical Personalities
cups, clear plastic	Sensing pH
dart guns (with hard plastic darts)	The Newtonian Shot
defrosting trays	I'm Melting! I'm Melting!
diethyl acetate (finger nail polish remover)	Circular Rainbows
diffraction grating, any	Bright Lights
digital multimeter (DMM)	The Lemon Electric
dollar bill	Reaction Time
Dominoes	Chain Reaction
dowel, wood	Motor Madness
drinking straws	Reckoning Latitude
dynamics cart and track	Putting the Force Before the Cart, An Uphill Climb
electrophorus	Electroscopia
electroscope	Electroscopia
elements and compounds	Chemical Personalities
epsom salt	Mystery Powders
Erlenmeyer flasks, 250-mL	Bubble Round-Up, Upset Stomach
ethanol	Oleic Acid Pancake, Circular Rainbows, Name that Recyclable
evaporating dishes	Chemical Personalities
eyedropper	Oleic Acid Pancake, Chemical Personalities, Mystery Powders
flashlight, small bright (LED)	Why the Sky is Blue
food coloring	Hot Spot Volcanoes, Solar Power II
force sensors	Force Mirror
forceps	Crystal Growth
galvanometer	Generator Activator
gas discharge tubes	Bright Lights
generator, hand-crank	Be the Battery, Motor Madness, Generator Activator

Geyser tube (or equivalent)	The Fountain of Fizz
glass slides	Crystal Growth
globe	Why Winds Wind Around Earth
gloves	Pure Sweetness, Upset Stomach
graduated cylinders, 100-mL	Thickness of a BB Pancake, Solar Power II
graduated cylinders, 10-mL	Oleic Acid Pancake
grapefruit juice	Sensing pH
hair dryer	Electroscopia
hardness set	What's That Mineral?
heat lamp and base	Canned Heat: Heating Up
heat-proof glove or potholder	Bright Lights
hex nuts	Putting the Force Before the Cart
hot plate	Chemical Personalities, Crystal Growth
hydrochloric acid (HCl)	Bright Lights, Chemical Personalities, What's That Mineral?, What's That Rock?
ice cubes	I'm Melting! I'm Melting!, Chemical Personalities
interface device (for sensors)	Sonic Ranger, Putting the Force Before the Cart, Force Mirror
iodine crystals	Chemical Personalities
iron filings	Magnetic Personality
isoamyl alcohol	Smells Great!
isobutanol	Smells Great!
isopentenol	Smells Great!
jars, baby food	Pure Sweetness
jars, glass	Solar Power I, Indoor Clouds
kitchen knife	Pure Sweetness
laser pointer, any	A Sweet Mirage
Laser Viewing Tank (or equivalent)	Why the Sky is Blue
lemons	The Lemon Electric
light beam viewing tank	A Sweet Mirage
light bulbs, clear 100-watt	Solar Power I
light source (OHP or equivalent)	Blackout
liter container	Temperature Mix
lithium chloride powder	Bright Lights
luminol crystals	Chemical Personalities
lycopodium powder	Oleic Acid Pancake
magnafier (small, 60x — 100x)	Pixel Peeping
magnet, neodymium	Dropping the Ball
magnetic field projectual	Magnetic Personality
magnets, bar	Magnetic Personality, Motor Madness, Generator Activator
marble (playing)	Why Winds Wind Around Earth
mass blocks	Putting the Force Before the Cart
mass hanger	The Weight
masses, slotted	Walking the Plank, The Weight
matches	Charging Ahead
measuring cup	Indoor Clouds
Mentos candies	The Fountain of Fizz
metal tube (small)	Electroscopia

meterstick	Go! Go! Go!, Walking the Plank, An Uphill Climb, Dropping the Ball, Fork It Over, Solar Power I, Solar Power II, Sunballs
methanol	Chemical Personalities, Circular Rainbows, Smells Great!
micrometer	Thickness of a BB Pancake
microscope	Crystal Growth
microspatula	Chemical Personalities
mineral collection	What's That Mineral?
mirror, full length	Mirror rirroM
molecular modeling kit	Molecules by Acme
mortar and pestle	Upset Stomach
motion sensor	Sonic Ranger, Putting the Force Before the Cart
nails, galvanized	The Lemon Electric
nails (short)	Spiked Water
n-propanol	Smells Great!
nuts, 1/2-inch	Sugar Soft
octanol	Smells Great!
oleic acid	Oleic Acid Pancake
paint, flat black	Solar Power I
paper towels	Spiked Water, Canned Heat: Heating Up, Canned Heat: Cooling Down, I'm Melting! I'm Melting!, Why the Sky is Blue
paper, butcher	Go! Go! Go!
paper, circular filter	Circular Rainbows
paper, graph	Go! Go! Go!, The Weight, Canned Heat: Heating Up, Canned Heat: Cooling Down, Ohm, Ohm on the Range, Sugar Soft, Tracking Mars
paperclip	Putting the Force Before the Cart
pebbles	Stream Study
pencils, colored	Over and Under
pennies	The Lemon Electric (optional)
pens, black felt-tip	I'm Melting! I'm Melting!, Circular Rainbows, Why Winds Wind Around Earth
petri dishes	Crystal Growth
phenolphthalein	Mystery Powders
PhET simulations	Pushing Things Around, Worlds of Wonder, Pool Cubes: Density, Pool Cubes: Buoyancy, Bouncing Off the Walls, Faraday's Electromagnetic Lab, Water Waves in an Electric Sink, Pixel Peeping
photogate timers	Blowout (optional), Dropping the Ball
pie tins, small	Charging Ahead
pipets, 9-inch plastic	Sugar Soft, Upset Stomach
plaster of Paris	Mystery Powders
plastic wrap	Solar Power II
plates (plastic or paper)	The Lemon Electric
playing cubes (or painted sugar cubes)	Get a Half-Life
plumb line and bob	Reckoning Latitude
polarizers, large sheet	Blackout
polarizers, small (slide-mounted)	Blackout
poster board	Why Winds Wind Around Earth
potassium aluminum sulfate (alum)	Crystal Growth
potassium chloride powder	Bright Lights

power resistors	Ohm, Ohm on the Range
power supply, variable DC (0–6 V)	Ohm, Ohm on the Range
propionic acid	Smells Great!
protractor	An Uphill Climb, Over and Under, Walking on Water, Reckoning Latitude, Tracking Mars
pulley	Putting the Force Before the Cart
pushpins	Reckoning Latitude
recyclable plastics	Name that Recyclable
red cabbage	Sensing pH
ring clamp	Electric Magnetism
ring stand	Things That Go Bump, Bright Lights, Chemical Personalities, Bubble Round-Up, Tubular Rust
rock samples	What's That Rock?
rod clamps	Walking the Plank, An Uphill Climb, The Weight, Motor Madness
rubber bands	Force Mirror, Motor Madness, Tubular Rust, Solar Power II
rubber stoppers, drilled	Bubble Round-Up, Solar Power I
ruler, centimeter	Mirror rorriM, Thickness of a BB Pancake, Sugar Soft, Tubular Rust, Top This, Walking on Water, Tracking Mars
safety goggles	Egg Toss, Mystery Powders, Salt and Sand, Bubble Round-Up, Pure Sweetness, Sensing pH, Upset Stomach
salicylic acid	Smells Great!
salt, table	Mystery Powders, Salt and Sand, Sensing pH
salt shaker (or sieve)	Why Winds Wind Around Earth
sand	Salt and Sand, Stream Study
scattering agent	Why the Sky is Blue
scoop (or large spoon)	Stream Study
silk cloth square	Electroscopia
soda pop	The Fountain of Fizz
sodium carbonate solution	Chemical Personalities
sodium chloride	Crystal Growth
sodium chloride powder	Bright Lights
sodium chloride solution	Chemical Personalities
sodium hydroxide	Mystery Powders
sodium nitrate	Crystal Growth
sodium perborate crystals	Chemical Personalities
sodium sulfate solution	Chemical Personalities
soil samples	Soil and Water
solder, lead-free	Motor Madness
spatula	Mystery Powders
spectroscope	Bright Lights
sponge (or Styrofoam block)	Hot Spot Volcanoes
spring scales	Walking the Plank, Force Mirror, An Uphill Climb, The Weight
steel wool	Chemical Personalities, Tubular Rust
stereo audio device	Sound Off
stirring rod	Why the Sky is Blue, Chemical Personalities
stopwatch	Go! Go! Go!, The Fountain of Fizz, Chain Reaction, Soil and Water
strainer	Sensing pH

wires, connecting	Ohm, Ohm on the Range, Batteries and Bulbs, Electric Magnetism, Motor Madness, Generator Activator
wood block	Putting the Force Before the Cart, Things That Go Bump, Motor Madness
wool cloth squares	Electroscopia

A Partial List of Vendors and Their URLs

Arbor Scientific · www.arborsci.com	Pasco Scientific · www.pasco.com
Sargent-Welch · www.sargentwelch.com	Science Kit · www.sciencekit.com
Vernier · www.vernier.com	

Name _____ Section _____ Date _____

CONCEPTUAL PHYSICAL SCIENCE

Experiment

Patterns of Motion and Equilibrium

The Equilibrium Rule

Walking the Plank

Purpose
To measure and interpret the forces acting on an object in equilibrium

Apparatus
meterstick
2 support rods
2 rod clamps
2 spring scales (5- or 10-newton capacity)
2 20-cm lengths of string

2 table clamps
2 crossbars (short rods)
2 collar hooks
slotted masses (two 200 g and one 500 g)
small spirit level (optional)

Discussion
Consider sign painters Burl and Paul, who work on a scaffold (a plank of wood suspended by ropes at both ends). They might wonder about the tension in the ropes that support their plank. They are in a state of equilibrium, but how are the rope tensions related to their weights and the weight of the scaffold? While the weights of Burl and Paul don't change, the tensions in the ropes do change when either of them moves along the plank. In this activity, you'll use a meterstick for such a scaffold. You'll measure the forces acting on the scaffold when it is in various arrangements, and interpret the forces that determine the condition of equilibrium.

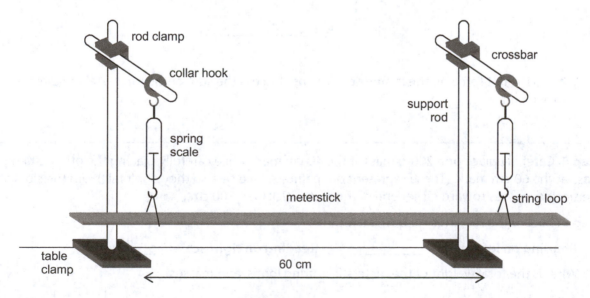

Figure 1

WALKING THE PLANK

Procedure

Step 1: Calibrate both spring scales so that when held vertically and carrying no load, each reads zero.

Step 2: Arrange the apparatus as shown in Figure 1.

a. Position the table clamps so that the support rods are about 60 cm apart.

b. Attach the crossbars to the support rods using the clamps. Hang a spring scale from each of the crossbars using the collar hooks.

c. Tie the ends of one 20-cm length of string together to create a loop. Hang the loop from one of the spring scales. Repeat for the other spring scale.

d. Suspend the meterstick (centimeter scale up) from the loops of string. Balance the arrangement so that the 50-cm mark is centered between the string loops and the meterstick is level. (Use a spirit level—if one is available—to check the meterstick.) This structure is a model of our painters' scaffold.

e. Adjust the meterstick so that the readings on both spring scales are the same (or very nearly the same). Move the meterstick left or right or adjust the level if necessary.

Step 3: Record the readings on both scales.

Reading on left scale:_____ **Reading on right scale:**_____

a. Add those readings and record the result. This is the total weight of the meterstick and string loops.

_____

b. Complete the diagram of the meterstick with the forces acting on it. The force *L* is the upward force on the left, *R* is the force on the right, and *W* is the downward force of weight.

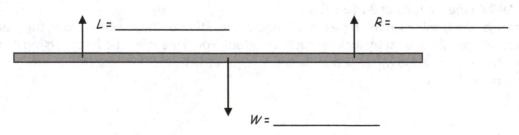

c. What is the net force on the meterstick? The net force is the sum of the forces, taking direction into account.

_____

Step 4: Carefully place one 200-g mass at the 40-cm mark while carefully placing the other 200-g mass at the 60-cm mark. (These represent our painters; take care so they don't fall!) Aim the slots of the slotted masses toward either end of the meterstick (0 or 100 cm).

Step 5: Record the readings for both scales.

Reading on left scale:_____ **Reading on right scale:**_____

a. What is the total weight of the meterstick, string loops, and masses?

_____

b. Sketch a diagram of the meterstick with all the forces acting on it. Include the numerical values of each force in your diagram.

c. What is the net force on the meterstick?

_____

Step 6: Move the mass at the 40-cm to the 70-cm mark. Keep the other mass at the 60-cm mark. The scaffold is still in equilibrium, even though the load is not evenly distributed.

Step 7: Record the readings for both scales.

Reading of left scale:_____   **Reading of right scale:**_____

a. What is the total weight of the meterstick, string loops, and masses?

_____

b. Sketch a diagram of the meterstick with all the forces acting on it. Include the numerical values of each force in your diagram.

c. What is the net force on the meterstick?

_____

d. Review your findings so far. What would you say is the condition for equilibrium, a condition that was met in all the arrangements investigated so far?

_____

_____

_____

_____

_____

Step 8: Suppose two painters with different weights used a scaffold. Simulate this by using a 500-g mass and a 200-g mass. Carefully stack the two masses at the 50-cm mark and read the scales.

    Reading of left scale:_____   **Reading of right scale:**_____

Step 9: Carefully place the 200-g mass at the 60-cm mark and the 500-g mass at the 40-cm mark, but do not read the scales yet.

What will the scale readings add to?

_____

_____

Step 10: Read **only** the left scale and record the reading.

    Reading on left scale:_____

Predict the approximate value of the reading on the right scale and record your prediction.

    Prediction on right scale:_____

Step 11: Read the right scale and record the reading.

    Reading on right scale:_____

How did the reading compare to your prediction?

_____

Step 12: Move the 200-g mass until both spring scales have the same reading. Record the location of the 200-g mass.

    Position of the 200-g mass:_____

The 500-g mass is 10 cm from the center of the meterstick. How far is the 200-g mass from the center of the meterstick?

_____

Summing Up

1. Can the meterstick platform be in equilibrium if the two upward support forces are equal to each other? If so, give an example from your observations.

_____

_____

_____

2. Can the meterstick platform be in equilibrium if the two upward support forces are unequal? If so, give an example from your observations.

_____

_____

_____

3. Would the platform be in equilibrium if a 500-g mass were at the 30-cm mark and a 200-g mass were at the 60-cm mark? Explain.

_____

_____

_____

4. Suppose the 500-g mass were placed at the 30-cm mark. Where could you place the 200-g mass so that both spring scales would have the same reading? Explain your answer.

_____

_____

_____

5. Could you use the same masses to get both scales to have the same reading if the 500-g mass were placed at the 20-cm mark? If so, where should the 200-g mass be placed? If not, why not?

_____

_____

_____

This experiment centers on an experience with sign painters Paul Hewitt and Burl Grey (see page 20 in the textbook), which led to Paul studying physics. Thanks to Paul's friend, Howie Brand, for suggesting this experiment.

CONCEPTUAL PHYSICAL SCIENCE | **Experiment**

Patterns of Motion and Equilibrium | The Basics of Graphing Motion

Go! Go! Go!

Purpose
To plot a graph that represents the motion of an object

Apparatus
constant-velocity toy car
butcher paper (or nonperforated paper towel, or several sheets of paper taped together)
access to tape (adhesive or masking)
stopwatch   meterstick   graph paper

Discussion
Sometimes the relationship between two quantities is easy to see. And sometimes the relationship is harder to see. A graph of the two quantities often reveals the nature of the relationship. In this experiment, you will plot a graph that represents the motion of a real object.

Procedure
You will observe the motion of the toy car. By keeping track of its position relative to time, you can make a graph that represents its motion. To do this, you will let the car run along a length of butcher paper. At 1-second intervals, you will
mark the position of the car. This will result in several ordered pairs of data—positions at corresponding times. You can then plot these ordered pairs to make a graph representing the motion of the car.

Step 1: Fasten the butcher paper to the top of your table with tape. It should be as flat as possible—no hills or ripples.

Step 2: If the speed of the toy car is adjustable, set it to the "slow" setting.

Step 3: Aim the car so that it will run the length of your table. Turn it on, and give it a few trial runs to check the alignment.

Step 4: Practice using the stopwatch. For this experiment, the stopwatch operator needs to call out something like "Go!" at each 1-second interval. Try it to get a sense of the 1-second rhythm.

Step 5: Practice the task.

a. Aim the car to drive down the length of the butcher paper and let it go.

b. *After* it starts, the stopwatch operator will start the stopwatch and say, "Go!"

c. Another person in the group should practice marking the location of the front or back of the car on the butcher paper every time the watch operator says, "Go!" For the practice run, simply touch the eraser of the pencil to the butcher paper at the appropriate points.

d. The stopwatch operator continues to call out, "Go!" (*not* "1, 2, 3 . . . ") once each second, and the marker continues to practice marking the location of the car until the car reaches the end of the butcher paper or table. Take care to keep the car from running off the table!

GO! GO! GO!

Step 6: Perform the task.

a. Aim the car to drive down the length of the butcher paper and **let it go**. At this point, **no** marks have been made on the butcher paper. None! The car is moving and no marks have been made.

b. **After** the car begins moving, the stopwatch operator will start the stopwatch and say, "Go!"

c. Another person in the group will mark the location of the front or back of the car on the butcher paper every time the watch operator says, "Go!" **No marks are to be made on the paper until the car is moving. Resist your urge to mark the location of the car when it's at rest!**

d. The watch operator continues to call out, "Go!" (**not** "1, 2, 3 . . . ") once each second, and the marker continues to mark the location of the car until the car reaches the end of the butcher paper or table. Take care to keep the car from running off the table!

Step 7: Label the marked points. The first mark is labeled "0," the second is labeled "1," the third is "2," and so on. These labels represent the times at which the marks were made.

Step 8: Measure the distance—in centimeters—of each point from the point labeled "0." (The "0" point is 0 cm from itself.) Record the distances on the data table. **Don't worry if you don't have as many data points as there are spaces available on the data table.**

Data Table

Time t (s)	0	1	2	3	4									
Position x (cm)	0													

Step 9: Make a plot of position vs. time on the graph paper. Title the graph "Position vs. Time." Make the horizontal axis time and the vertical axis position. Label the horizontal axis with the quantity's symbol and the units of measure: "t (s)". Label the vertical axis in a similar manner. Make a scale on both axes starting at 0 and extending far enough so that all your data will fit within the graph. Don't necessarily make each square equal to 1 second or 1 centimeter. Make the scale so the data will fill the maximum area of the graph.

We could just as easily make a graph of time vs. position. But we prefer position vs. time for a few reasons. In this experiment, time is what we call an *independent variable*. That is, no matter how fast or slow our car was, we always marked its position at equal time intervals. **We were in charge of the time intervals; the *car* was "in charge" of the change in position it made in each interval. But the change in position of the car in each interval depended on the time interval we chose. So we call position the *dependent variable*. We generally arrange a graph so that the horizontal axis represents the independent variable and the vertical axis represents the dependent variable.

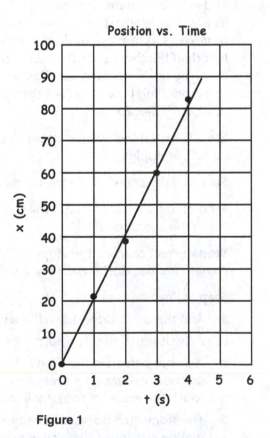

Figure 1

Also, the slope of a position vs. time graph tells us more than the slope of a time vs. position, as we will see later.

Step 10: Draw a line of best fit. In this case, the line of best fit should be a single, straight line. Use a ruler or straight edge; place it across your data points so that your line will pass as closely as possible to all your data points. The line may pass above some points and below others. Don't simply draw a line connecting the first point to the last point. An example is shown in Figure 1.

Step 11: Determine the slope of the line. **Slope** is often referred to as "rise over run." To determine the slope of your line, proceed as follows.

a. Pick two convenient points on your line. They should be pretty far from each other. Convenient points are points where grid lines intersect on the graph paper.

b. Extend a horizontal line to the right of the lower convenient point, and extend a vertical line downward from the upper convenient point until you have a triangle as shown in Figure 2. It will be a right triangle, because the horizontal and vertical lines meet at a right angle.

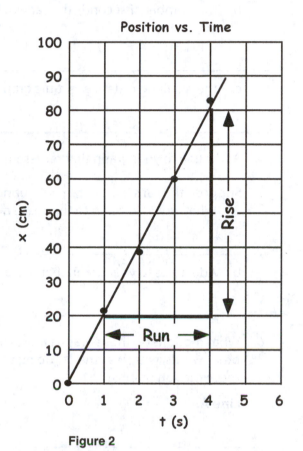

c. Find the length of the horizontal line on your graph. This is the "run." Don't use a ruler; the length must be expressed in units of the quantity on the horizontal axis—in this case, seconds of time.

   Run: _____ s

d. Measure the length of the vertical line. This is the "rise." Don't use a ruler; the length must be expressed in units of the quantity on the vertical axis—in this case, centimeters of distance.

   Rise: _____ cm

Figure 2

e. Calculate the slope by dividing the rise by the run. **Show your calculation** (and include appropriate units) in the space below.

   Slope: _____ cm/s

Summing Up

1. Suppose a faster car were used in this experiment. What would have been different about
   a. the distance between the marks on the butcher paper?

   _____

   b. the number of seconds the car spent on the butcher paper before reaching the edge?

   _____

   c. the resulting distance vs. time graph? (How would the slope have been different?)

   _____

GO! GO! GO!

2. Add a line to your graph that represents a faster car. Label it appropriately.

3. Suppose a slower car were used in this experiment. What would have been different about
   a. the distance between the marks on the butcher paper?

   _____

   b. the number of seconds the car would have spent on the butcher paper before reaching the edge?

   _____

   c. the resulting distance vs. time graph? (How would the slope have been different?)

   _____

4. Add a line to your graph that represents a slower car. Label it appropriately.

5. Suppose the car's battery ran out during the run so that the car slowly came to a stop.
   a. What would happen to the space between marks as the car slowed down?

   _____

   b. Add a line to your graph that represents a car slowing down. Label it appropriately.

   _____

6. What motions do these graphs represent? In other words, what was the car doing to generate these motion graphs?

   Line A.

   _____

   _____

   _____

   Line B.

   _____

   _____

   _____

Position vs. Time

CONCEPTUAL PHYSICAL SCIENCE	**Activity**

Patterns of Motion and Equilibrium   **Graphing Motion in Real Time**

Sonic Ranger

Purpose
To produce and interpret real-time graphs of your position using a motion sensor

Apparatus
motion sensor (sonic ranging device)
interface device and connectors
motion sensor software
computer

Discussion
Graphs can be used to represent motion. For example, if you track the position of an object as time goes by, you can make a plot of position vs. time. In this activity, the sonic ranger will track your position, and the computer will draw a position vs. time graph of your motion. The sonic ranger sends out a pulse of high-frequency sound and then listens for the echo. By keeping track of how much time goes by between each pulse and corresponding echo, the ranger determines how far you are from it. (Bats use this technique to navigate in the dark.) By continually sending pulses and listening for echoes, the sonic ranger tracks your position over a period of time. This information is fed into the computer, and the software generates a position vs. time graph.

Procedure
Your instructor will provide a computer with a sonic ranging program installed. Check to see that the sonic ranger is properly connected and operating reliably. Position the sonic ranger so that its beam is about chest high and aimed horizontally. (*Note*: Sometimes these devices do not operate reliably on top of computer monitors.)

The sonic ranger should be set to "long range" mode. The computer should be set to "graph position vs. time." Initiate the sonic ranger, and note how close and how far you can get before the readings become unreliable.

PART A: MOVE TO MATCH THE GRAPH
Generate real-time graphs of each motion depicted on the following pages, and write a description of each. ***Do not use any form of the term "acceleration" in any of your descriptions.*** Instead, use terms and phrases such as "rest," "constant speed," "speed up," "slow down," "toward the sensor," and "away from the sensor."

Study each graph below. When you are ready, initiate the sonic ranger and move so that your motion generates a similar graph. Then describe the required motion in words.

Example:

Position vs. Time Graph

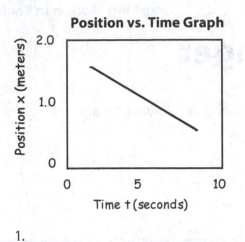

Description

Move toward the sensor at constant speed.

Make sure each person in the group can move to match this graph before going on to the next graph.

1.

Position vs. Time Graph

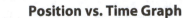

Description

2.

Position vs. Time Graph

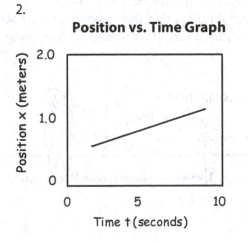

Description

3.

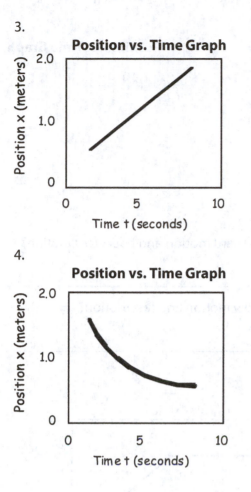

Position vs. Time Graph

Description

4.

Position vs. Time Graph

Description

PART B: MOVE TO MATCH THE WORDS
Walk to match each description of motion. Draw the resulting position vs. time graph.

5.

Description

Move toward the sensor at constant speed, stop and remain still for a second, then walk away from the sensor at constant speed.

Position vs. Time Graph

6.

Description

Move toward the sensor with decreasing speed, then just as you come to rest, move away from the detector with increasing speed.

Position vs. Time Graph

SONIC RANGER

7.

| **Description** | **Position vs. Time Graph** |

Move away from the sensor with decreasing speed until you come to a stop. Then move toward the sensor with decreasing speed until you come to a stop.

Summing Up

1. How does the graph show the difference between forward motion and backward motion?

_____

2. How does the graph show the difference between slow motion and fast motion?

_____

3. Study the graph of position vs. time shown below.

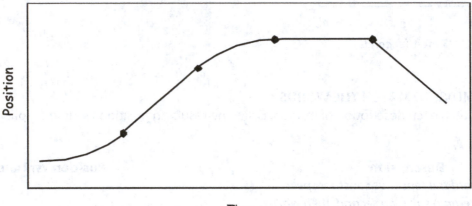

Label sections of the graph showing where the object is
- at rest
- moving forward at constant speed
- moving backward at constant speed
- speeding up
- slowing down

Going Further

Some motion graphing software includes an option for "graph matching." The software displays a plot on the screen, and you must move so that your plot matches the graph on the screen. At the end of the run, the score of your match is displayed. Explore this option and see how high you can score.

CONCEPTUAL PHYSICAL SCIENCE	Activity

Newton's Laws of Motion   **Force, Mass, and Acceleration Simulation**

Pushing Things Around

Purpose
To develop an understanding of Newton's second law of motion through the use of a computer simulation capable of showing the relationship among force, mass, and acceleration

Apparatus
computer
PhET sim "Force and Motion: Basics" (available at http://phet.colorado.edu)

Discussion
There is an important relationship between force and motion. Using the simulation (sim), we will investigate the basics of this relationship.

Procedure
SETUP
Turn the computer on, log on, and allow it to complete its start-up cycle. Find and start the PhET simulation "Forces and Motion: **Basics.**" The sim may be loaded on the computer, or you may need to access the PhET website via the Internet. Ask your instructor for assistance if needed. *Note*: There is also a sim called "Forces and Motion." That is *not* the sim for this activity.

PART A: TUG OF WAR
The sim should open in the Tug of War tab. If it does not, click the tab to activate Tug of War.

In this sim, a wagon (W) can be pulled by the players of two teams. Both Blue Team and Red Team have four players, one adult (A), a teenager (T), and two children (C). The adult is the largest player; the children are the smallest players.

Step 1: Even Steven
a. If you arrange identical teams in the Tug of War, you'll get a tie. The net force is zero. The following are examples of identical teams.

A---W---A (Adult vs. Adult)

C-T---W---C-T

A-T-C-C---W---C-C-T-A

C-A-T---W---A-C-T

b. Describe three distinct ways to arrange the teams so that you get a tie even though the teams have **different numbers** of players. These are non-identical, balanced teams. (To show distinct arrangements, don't simply switch players on each side. For example, **A-C---W---T** and **T---W---A-C** are **not** distinct. (Nor do they produce a tie.)

i. _____

ii. _____

iii. _____

c. Complete this statement, which relates part of Newton's first law (the law of inertia) to these observations:

A body _____ _____ **will remain** _____ _____

unless acted on by a/an _____ **force.**

Step 2: The Electric Slide

a. Arrange an unbalanced tug of war by placing only one player on the rope. One team will have an active player; the other team will be completely unmanned. Click the on-screen Go button, wait two full seconds, then click the on-screen Pause button. At this point, the wagon was moving and the one-player team was winning.

b. While the action remains paused, add player(s) to the other team so that the two teams are balanced (the net force is zero).

c. Prediction. While the action remains paused, predict what will happen when the sim is unpaused and action resumes with the newly balanced teams.

_____

_____

d. Observation. Now unpause the sim and record your observation of what happens.

_____

_____

e. Explain. Use the law of inertia to explain why this happens.

_____

_____

PART B: MOTION
Click the Motion tab to select the Motion activities.

Step 1: PhET Girl vs. Massive Stack
a. In the on-screen control panel, the Force checkbox should already be selected. Click the Speed checkbox to activate the speedometer. Set the PhET Girl on the skateboard. Use maximum Applied Force to get her up to speed as rapidly as possible.

    i. What happens when she reaches maximum speed? (Discuss the pusher, the motion of the girl, and the grayed-out—unavailable—section of the Applied Force control.)

_____

_____

    ii. Now apply a maximum force in the **opposite** direction until the pusher falls again. What happens during this push?

_____

_____

b. Take the girl off the skateboard, and load the skateboard with two crates and the refrigerator. Repeat the maximum force one way, then the other way, as was done with the PhET Girl.

    What was different this time, and why?

_____

_____

PART C: ACCELERATION LAB
Click the Acceleration Lab tab. Activate the Speed and Acceleration displays. Set the Friction to None (Ice).

Step 1: Sloshy
a. Set the translucent/transparent bucket on the track and push it for maximum acceleration. Record observations of the speedometer, the accelerometer, and the water in the bucket.

_____

_____

_____

b. Remove the bucket to stop the action. Now stack a crate, a refrigerator, and the bucket. Push again for maximum acceleration. Describe the differences compared to part a.

_____

_____

_____

Step 2: Numerical-ish
Click the on-screen Reset All button. Set Friction to None. In the control panel, activate Forces: Values, Masses, Speed, and Acceleration.

a. Click the on-screen Pause button. Set one 50-kg crate on the track. Set the Applied Force to 200 N. (If the slider won't let you select 200 N exactly, type 200 N into the Applied Force value space). Unpause the sim for a second or two so that the acceleration registers on the accelerometer. Then pause the sim. For the purposes of the questions below, we will consider this configuration to be the Original Arrangement.

   i. From the Original Arrangement, what **single** change could double the acceleration?

   _____

   ii. From the Original Arrangement, how could you **halve** the acceleration *without* changing the Applied Force?

   _____

   _____

b. Remove all items from the track to stop the action.

   i. Click the on-screen Pause button. Add two 50-kg crates to the track. Set the Applied Force to 250 N. Unpause the sim momentarily to register the acceleration.

   ii. Make two changes to the arrangement so that the acceleration will **quadruple** when the sim is unpaused. Describe the changes.

   _____

   _____

Summing Up
Describe the relationship among force, mass, and acceleration.

_____

_____

Name _____ Section _____ Date _____

CONCEPTUAL PHYSICAL SCIENCE	Activity

Newton's Laws of Motion   **Force, Mass, and Acceleration**

Putting the Force Before the Cart

Purpose

To observe and interpret the motion of several objects acted on by varying forces to learn the relationship among force, mass, and acceleration

Apparatus

dynamics cart and track
mass blocks
string (about 1 meter)
pulley
wood block (or end stop)

paper clip
4 hex nuts (or equivalent)
computer with motion-graphing software
motion sensor and interface device

Discussion

Some of the most fundamental laws of motion eluded the best minds in science for centuries. One reason for this is that when objects are pushed or pulled, there are usually several forces acting at once. To understand the nature of force and motion, it is necessary to observe the effect of a single, unbalanced force acting on an object. In this activity, you will do just that. You will also vary the amount of force acting on the object. You will then vary the mass of the object being acted upon by a force. Careful observations will lead you to an understanding of the relationship among force, mass, and acceleration—a relationship that Galileo missed, and Newton got!

Procedure

Step 1: Connect the motion sensor to the computer (using the interface device). If the motion sensor has a range selector, choose the "short range" setting. Activate the motion-graphing software.

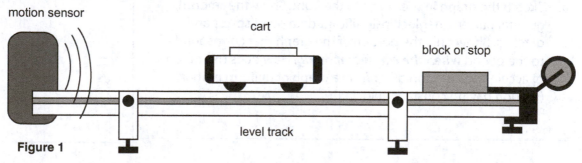

Figure 1

Step 2: Arrange the apparatus as shown in Figure 1.

a. Make sure the track is level. The cart should be able to coast equally in either direction along the track. If the cart "prefers" to roll in one direction, adjust the track accordingly.

b. The pulley clamp should be secure on the track, and the pulley should be able to spin freely.

c. Set the wood block on the track (or employ some other stopping mechanism) so the cart cannot roll into the pulley.

d. Arrange the string so that it is attached to the cart at one end and to the paper clip at the other end, as shown in Figure 2. The length of the string is such that when the cart is stopped at the wood block, the paper clip does not touch the ground. If the paper clip touches the ground, shorten the string.

Step 3: Test the sensor and software.

a. Place the cart near the middle of the track.

b. Activate the motion sensor and the graphing software.

c. Move the cart back and forth with your hand. The computer should show a graph that corresponds to the motion of the cart. If it does not, adjust the aim of the sensor and check the connecting wires.

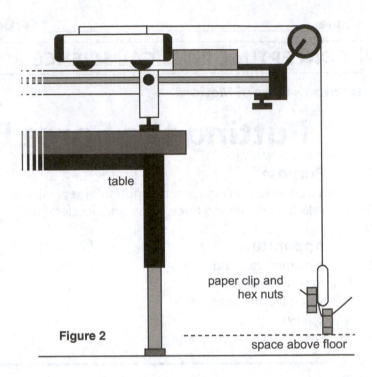

table

paper clip and hex nuts

Figure 2

space above floor

If the problems persist, ask your instructor for assistance.

PART A: VARY THE FORCE

Step 4: Check to see that there are two hex nuts attached to the paper clip. Pull the cart back so the paper clip is just below the pulley wheel. Make note of the cart's starting position on the track.

Step 5: Clear the computer of any previous trials, and activate the motion sensor.

Step 6: When the motion sensor begins sampling, release the cart and allow it to move along the track until it is stopped by the wood block.

Step 7: Deactivate the motion sensor. If something went wrong during the trial, simply reset the cart, string, and software, and repeat the trial so that you have a reliable result.

a. Sketch the graph in the space to the right. Show the smooth, general pattern; neglect insignificant data point spikes and glitches. Show only the portion of the graph that corresponds to the period **when the cart was moving**. How does this graph of accelerated motion differ from a graph of uniform motion (motion having constant velocity)?

_____

_____

_____

b. What do you think will happen to the acceleration if twice as much force is used to pull the cart?

_____

_____

Step 8: Add the two remaining hex nuts to the paper clip (for a total of four). This will double the force pulling the cart.

Step 9: Set the cart in place for a second trial, starting from the same position on the track. Prepare the software to add a second trial to the one already recorded. Activate the sensor. When the sensor begins sampling, release the cart. When the cart is stopped, deactivate the sensor.

How does the acceleration caused by the doubled force compare with the original acceleration (from Step 6)? Did your observation confirm or contradict your prediction?

_____

_____

PART B: VARY THE MASS

Step 10: Determine the mass of your cart and record it here: _____

Step 11: Clear the computer of any previous trials.

Step 12: Attach four hex nuts to the paper clip. Set the cart in place. Activate the sensor. When the sensor begins sampling, release the cart. When the cart is stopped, deactivate the sensor.

Step 13: Add a mass block or blocks to the cart so that the mass is doubled. For example, if the cart has a mass of 500 g, add 500 g of mass blocks to it. Do not change the hex nut configuration.

What do you think will happen to the acceleration if the same force is used to pull a cart having twice as much mass?

_____

_____

Step 14: Set the cart in place for a second trial. Prepare the software to add a second trial to the one already recorded. Activate the sensor. When the clicking begins, release the cart. When the cart is stopped, deactivate the sensor.

How does the acceleration of the doubled mass compare with the original acceleration (from Step 12)? Did your observation confirm or contradict your prediction?

_____

_____

Summing Up

1. How does the acceleration of the cart depend on the force pulling it?

    ___ **Greater force results in greater acceleration. In other words, acceleration is directly proportional to force.**

    ___ **Greater force results in lesser acceleration. In other words, acceleration is inversely proportional to force.**

    ___ **Greater force results in the same acceleration. In other words, acceleration is independent of force.**

2. How does the acceleration of the cart depend on the mass of the cart?

    ___ **Greater cart mass results in greater acceleration. In other words, acceleration is directly proportional to mass.**

    ___ **Greater cart mass results in lesser acceleration. In other words, acceleration is inversely proportional to mass.**

    ___ **Greater cart mass results in the same acceleration. In other words, acceleration is independent of mass.**

3. Complete the statement:

    The acceleration of an object is _____ proportional to the net

    force acting on it and _____ proportional to the mass of the

    object.

4. Which mathematical expression is most consistent with your observations?

    a. $a = F \cdot m$ **b.** $a = F / m$ **c.** $a = m / F$

5. Examine the position vs. time graphs plotted in the diagram to the right. Suppose that plot B represents an empty cart pulled by two hex nuts. If the mass of the cart were doubled and four hex nuts were used to pull the cart, which plot would best represent the result? Explain.

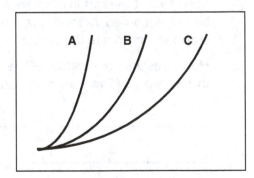

_____

_____

_____

_____

CONCEPTUAL PHYSICAL SCIENCE	Activity

Newton's Laws of Motion **Force and Motion Puzzle**

The Newtonian Shot

Purpose
To explore the role of Newton's laws and gravity as they determine the outcome of a race between two darts fired from two dart guns

Apparatus
2 spring-loaded dart guns that shoot hard-stick, suction-cup darts
2 darts (for the dart guns)
metal ball (steel or lead, approximately 1 inch in diameter)
wood or cork ball (same size as metal ball)

Discussion
Two darts are to be fired at the same time from two identical dart guns. Both will be fired from the same height and directed straight down toward the ground. Both darts have been modified. One is attached to a heavy metal ball, and the other is attached to a light wood or cork ball.

Procedure
Step 1: The Possibilities. Which ball will hit the ground first? Before you decide, list three possible outcomes and write arguments for each of them. Doing so means you will have to write arguments for two outcomes that you don't believe will occur.

a. Outcome 1: _____

_____ Supporting argument: _____

_____

b. Outcome 2: _____

_____ Supporting argument: _____

_____

c. Outcome 3: _____

_____ Supporting argument: _____

_____

Step 2: Your Prediction. Which dart will hit the ground first?

_____

Step 3: Observation. Allow your instructor to complete the demonstration by launching the two darts. Add to the illustration below. Sketch a snapshot of the moment when the "winning" dart hits the ground, showing the location of both darts at that instant.

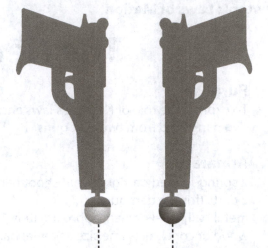

Summing Up

1. Which dart hit first?

_____

2. What is the reason for the outcome observed? Be specific and complete!

_____

_____

_____

_____

_____

CONCEPTUAL PHYSICAL SCIENCE	Activity

Newton's Laws of Motion   **Quantitative Observation of Force Pairs**

The Force Mirror

Purpose
To compare the sizes of "action" and "reaction" forces when two objects interact

Apparatus
2 spring scales (preferably with somewhat different ranges)
2 #64 rubber bands (#64 rubber bands are 0.25 inch wide)
computer with software for graphing force sensor data
2 force sensors
interface device(s) for connecting sensors to the computer
table clamp
support rod

Discussion
Forces are interactions between two objects. The interactions can arise in many ways: friction, tension, drag, and many more. But no matter what the nature of the interaction, two objects are required. There are always two objects, and there are always two forces. The two forces involved in a particular interaction form a **force pair.** You will use spring scales and force sensors to investigate a few force pairs.

Procedure
PART A: SPRING SCALES
Step 1: Attach the table clamp to the table, and attach the support rod to the table clamp.

Step 2: Make sure the spring scales are calibrated so that they read 0 when no force is applied to them. If you need assistance, ask your instructor.

Step 3: Take hold of one spring scale while a lab partner holds the other spring scale. Attach a rubber band to the hooks of the spring scales to connect them.

Step 4: While you hold your spring scale in place, have your partner pull his or her spring scale until a low but easy-to-read force is achieved.

Compare your reading to your partner's reading (take care to make sure that neither scale is pulled to its full-range limit). Which statement best describes your finding?

____ **My reading is much greater than my partner's.**

____ **My reading is much less than my partner's.**

____ **My reading is about the same as my partner's.**

Step 5: Make a second observation. This time, let your partner hold his or her spring scale in place while you pull your spring scale. How, if at all, does this change the outcome in terms of the force readings?

_____

_____

Step 6: Try other values of force to see whether this pattern is consistent. Be careful to make sure you don't reach the limit of either spring scale.
Does the pattern hold across a range of force values?

Step 7: Try connecting one spring scale to the support rod and pulling the other spring scale (while the two are still connected by the rubber band). Does the pattern still hold under these conditions?

PART B: FORCE SENSORS

Step 1: Turn on the computer and let it complete its start-up cycle.

Step 2: Connect the force sensors to the computer using the interface device(s). If you need assistance in this or any of the following steps, ask your instructor.

Step 3: Activate the software that will graph the force sensor data.

Step 4: Use the software to configure the force sensors as follows:
a.   One force sensor should be configured "Push Positive."

b.   The other force sensor should be configured "Pull Positive."

Step 5: Configure the software to show a force vs. time graph of the data from **both** sensors. That is, there will be two plots on one set of axes.

Step 6: Connect the hooks of the force sensors with two rubber bands.

Step 7: Set the force sensors down so that there is no force on their hooks. Press the "zero" button on each sensor to calibrate them.

Step 8: Activate the software's sampling mode (for example, push the on-screen "Start" button).

Step 9: You and your partner may now pick up the force sensors and pull with varying amounts of force. Continue for about 20 seconds and then stop the sampling.

Step 10: Ask your instructor whether the resulting graph shows that your sensors are correctly connected and calibrated. If not, make appropriate corrections and try again. Otherwise, continue.

Step 11: Suppose the graph in Figure 1 shows **one** sensor's plot on the force vs. time graph. Based on your experience in this lab, draw the plot of the **other** sensor.

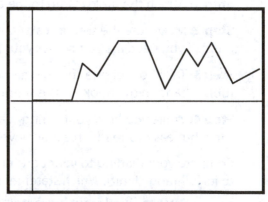

Step 12: Clear the previous force vs. time graph. Attach one force sensor to the support rod. Zero both sensors and connect them with rubber bands. Start sampling, and vary the force on the free sensor for about 20 seconds.

Figure 1. Force vs. Time

Summing Up
Complete the statement. Include the terms "magnitude" (size) and "direction" in your response.

When one object exerts a force on a second object, the second exerts a force on the first that is

CONCEPTUAL PHYSICAL SCIENCE	**Activity**

Newton's Laws of Motion   **A Summary of Newton's Three Laws**

Blowout!

Purpose
To observe a simple blowgun and describe its operation in terms of Newton's three laws of motion

Apparatus
1-inch-diameter tube (5 to 10 ft long)   marker pen
masking tape   support rods and clamps (to secure tube)
catch box

Optional Equipment
photogate timing system   balance

Discussion
The operation of a blowgun involves fundamental principles of physics. A force is applied to accelerate a mass to a relatively high speed. The mass travels some distance and is then brought to a stop. Force pairs are involved when the mass is speeding up and slowing down.

Procedure
Step 1: Use the support rods and clamps to hold the tube horizontally. See Figure 1. If the tube is long, or if the tube is PVC, you may need a center support to keep the tube from sagging.

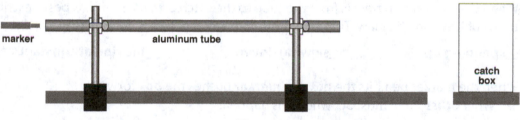

Figure 1

Step 2: Set the catch box some distance from the tube so that it will stop the marker pen when the pen comes out of the tube.

Step 3: If necessary, wrap the marker pen with enough masking tape so that the taped marker pen barely fits into the tube. See Figure 2. The pen must be free to move through the tube; you'll want a good seal to prevent air from blowing by the marker while it's in the tube.

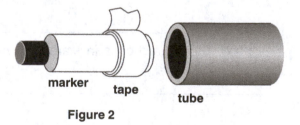

Figure 2

Step 4: Place the marker in the tube and clean the end of the tube so that it is sanitary.

Step 5: Blow the marker out of the tube. Make sure your mouth has a good seal on the tube, and blow with as much force as you can.

Summing Up

1. Which of Newton's laws best describes why the marker, initially at rest, requires a force to accelerate it? Why does this law apply to this situation?

_____

_____

2. Identify the force pair involved when the marker is propelled through the tube.

   The _____ pushes the _____ forward;

   the _____ pushes the _____ backward.

3. a. Which of Newton's laws determines the amount of acceleration that the marker pen experiences while being blown through the tube?

   _____

   b. Apply that law to describe the amount of acceleration the marker pen will experience.

   _____

   _____

   _____

4. When the marker pen travels from the tube to the catch box, its motion is best described in terms of Newton's first law. That is to say that the marker pen is, for the most part,

   __speeding up.   __slowing down.   __moving at constant velocity.

5. When the marker pen hits the box, the **marker** pushes the **box** forward.
   a. What **other** force must act when this happens?

   _____

   _____

   _____

   b. This interaction is an example of which law of motion?

   _____

Going Further

Step 1: Arrange the photogate system so that it can be used to determine the speed of the marker pen when it comes out of the tube. See Figure 3.

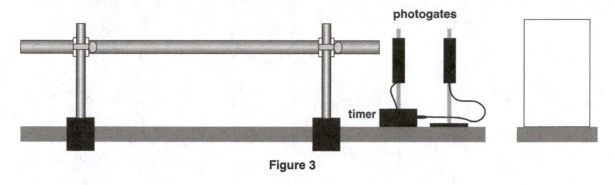

Figure 3

Describe the details of how the speed will be determined.

_____

_____

_____

Step 2: Repeat the demonstration and calculate the speed of the marker pen.

_____

_____

Step 3: Measure the length of the tube. The acceleration of the marker can be determined using the equation $a = v^2/2x$, where v is the speed of the marker and x is the length of the tube. Use the speed of the marker and the length of the tube to calculate the acceleration of the marker pen.

_____

_____

Step 4: Divide the acceleration by 9.8 m/s^2 to determine the number of g's experienced by the marker pen in the tube.

_____

_____

Step 5: Measure the mass of the marker pen. Use Newton's second law to calculate the force of the exhaled air on the marker pen when the pen is launched.

_____

_____

| **CONCEPTUAL PHYSICAL SCIENCE** | **Activity** |

Momentum and Energy   **Sticking and Bouncing in Collisions**

Things That Go Bump

Purpose
To investigate the difference between the forces applied in elastic collisions (when objects bounce) and the forces applied in inelastic collisions (when objects stick)

Apparatus
Bouncing Dart (Happy/Sad Ball-tipped demonstration pendulum)
2 support rods and necessary clamps
wooden block (such as a pair of friction blocks attached to each other)
ringstand with support rod
(Optional) Happy and Sad Balls (loose)
(Optional) heavy brick with a hard surface (or an equivalently stable object)

Discussion
When objects collide, they may bounce off each other or they may stick to each other. Happy and Sad Balls can be used to demonstrate the different interactions. In a collision, a Happy Ball bounces and a Sad Ball sticks (stops). Do these interactions involve different amounts of force? The implications range from steam engine paddlewheels to bullets used for crowd control by riot police!

Procedure
SETUP
Arrange the apparatus as shown. See Figure 1. The pendulum can swing freely.

Step 1: Identify which tip of the pendulum hammer is which. Place the heavy brick on the table at the pendulum's low point. Pull the pendulum up and release it. Note whether the hammer bounces or stops when it collides with the brick. Now arrange to have the other side of the hammer strike the brick, and observe the different behavior.

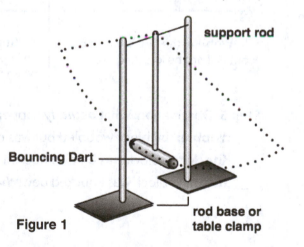

Figure 1

Step 2: Through trial and error, determine the minimum distance the Happy Ball (bouncy) side of the pendulum needs to swing through to **just barely** knock over the block. See Figure 2. Mark the pull-back distance (using the brick, for example), and repeat to make sure that this is the correct minimum distance.

THINGS THAT GO BUMP

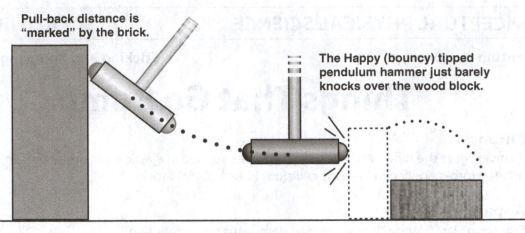

Pull-back distance is "marked" by the brick.

The Happy (bouncy) tipped pendulum hammer just barely knocks over the wood block.

Figure 2

Step 3: Reverse the pendulum so that the Sad Ball (sticking side) of the pendulum will strike the wood block, but do not carry out a collision just yet.

Step 4: Prediction. The Sad (sticking) side of the pendulum will be swung from the same distance. What will happen when it strikes the block?

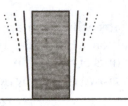

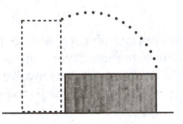

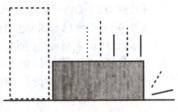

__ **Wobble:** the block will wobble but will not be knocked over.

__ **Knock:** the block will barely be knocked over, just as it was when struck by the bouncy tip.

__ **Slam:** the block will be knocked down hard.

Step 5: Observation. What *actually* happened when the Sad-tipped pendulum struck the block?

__ **Wobble:** the block wobbled but was not knocked over.

__ **Knock:** the block was barely knocked over, just as it was when struck by the Happy (bouncy) tip.

__ **Slam:** the block was knocked down *hard*.

Summing Up

1. Why does this outcome occur? Consider the following.

   If the pendulum hadn't hit anything, its momentum would not have changed at all. If it had had a momentum of "+1," its momentum would have continued as "+1." That is, $\Delta p = 0$.

   When the pendulum was brought to a stop, all its momentum was lost. If its initial momentum was "+1" and its final momentum was 0, then its change in momentum was "1" (technically, it was "−1," but let's just worry about the **size** of the change, not the sign.)

   What is the change in momentum when an object bounces—that is, when its momentum changes from "+1" to "−1"?

   _____

2. If the pendulum didn't hit anything, $\Delta p = 0$, and there would have been no force.

   When the pendulum was brought to a stop, $\Delta p = 1$, and some amount of force would have been required to stop the pendulum.

   How does the force involved in bouncing compare to the force involved in sticking (stopping)?

   _____

   _____

3. Consider two children, a pumpkin, and a wagon. Which of the following events will result in the greatest speed for the child in the wagon?

   A. Child in wagon throws pumpkin to other child.

   B. Child in wagon catches pumpkin thrown by other child.

   C. Child in wagon catches pumpkin and then throws it back.

   Defend your answer.

   _____

   _____

   _____

   _____

4. A paddlewheel was once a common method for turning a fast-moving stream of water into rotational motion. The water exerts a force on the paddles; this provides torque to turn the wheel.

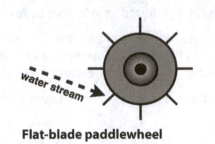

Flat-blade paddlewheel

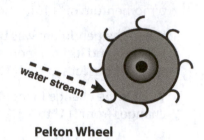

Pelton Wheel

Lester Pelton made a fortune with his redesigned paddlewheel (the Pelton wheel). What made his curved paddles so much more effective than flat paddles? Refer to the physics of this activity.

_____

_____

_____

_____

5. In the past, riot police commonly used rubber bullets instead of "live rounds" (metal bullets). Rubber bullets were preferred because they bounced off bodies rather than penetrating. But people hit by rubber bullets often sustained significant injuries. Rubber bullets have since been abandoned in favor of beanbags. Use the lessons of this activity to explain the advantages of beanbags over rubber bullets.

_____

_____

_____

_____

CONCEPTUAL PHYSICAL SCIENCE   **Experiment**

Momentum and Energy   **Work on an Inclined Plane**

An Uphill Climb

Purpose
To determine what advantage—if any—there is in using an inclined plane to move an object to a higher elevation

Apparatus
dynamics cart and track or board that can be inclined and secured at various angles
spring scale (capable of weighing the dynamics cart)

table	table clamp
support rod	rod clamp
meterstick	protractor

Discussion
Why are ramps used in lifting heavy objects? Does it make the task easier (requiring less force)? Does it make the movement shorter (requiring less distance)? Does it make the effort more efficient (requiring less work)? Perhaps it does several of these things; maybe it does none of them. You will learn more from this lab if you record your initial thoughts before making any measurements or calculations.

What advantages or disadvantages are there in using a ramp when lifting a heavy object?

_____

_____

In this experiment, the cart will act as the heavy object. Your task will be to move your cart a vertical distance of 20 cm above the tabletop. You will arrange a series of ramps (inclined planes) at different angles to accomplish this task. You will measure the force needed to move a cart up the incline. You will also measure the distance through which that force would be applied to finish the job. You will then calculate the work required to lift an object using an inclined plane. By the end of the experiment, you will be able to identify what an inclined plane can do for you in terms of force, distance, and work.

Procedure
PART A: SHALLOW INCLINE
Step 1: Arrange the apparatus as shown in Figure 1. The plane should be inclined at an angle between 20° and 30°. Check the angle with the protractor as shown in Figure 2.

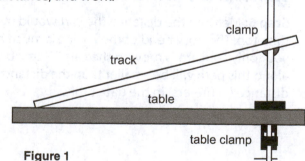

Figure 1

AN UPHILL CLIMB

Step 2: Measure the force needed to pull the cart along this incline using the spring scale as shown in Figure 3. The spring scale is held parallel to the inclined plane when the measurement is made. Because the force needed to move the cart at a constant speed is the same as the force needed to keep the cart at rest on the plane, measure the force when the cart is at rest. Record the force below, and transfer the value to the data table.

Shallow incline force:

F = _____ N

Step 3: Measure the distance the cart would travel along the inclined path to move from the tabletop to a distance 20 cm above the tabletop. See Figure 4. The path starts at the tabletop (even if the object being used as an inclined plane doesn't come all the way down to the tabletop). The path ends where the inclined plane is 20 cm above the tabletop. This distance will be greater than 20 cm for all inclined planes (unless the plane goes straight up). Convert the distance from centimeters to meters. Record the distance below, and transfer the value (in meters) to the data table.

Shallow incline distance:

d = _____ cm

= _____ m

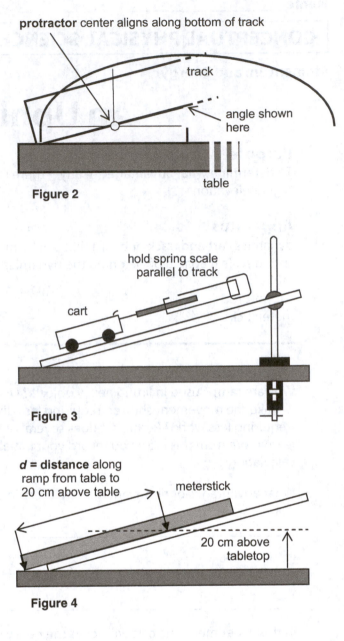

Figure 2

Figure 3

Figure 4

PART B: MEDIUM INCLINE

Step 4: Increase the angle of incline to a value between 40° and 50°.

Step 5: Measure the force needed to move the cart along this incline, and record it in the data table.

Step 6: Measure the distance the cart would travel along this path to move 20 cm above the tabletop. The upper end of the incline is much higher now than it was for the shallow incline, but your only concern is moving the cart 20 cm above the tabletop. The distance the cart would travel along this path will be smaller than the distance it would travel along the shallow incline. Record the distance (in meters) in the data table.

PART C: STEEP INCLINE

Step 7: Increase the angle of incline to a value between 60° and 70°.

Step 8: Measure the force needed to move the cart along this incline, and record it in the data table.

Step 9: Measure the distance the cart would travel along this path to move 20 cm above the tabletop. Record the distance (in meters) in the data table.

PART D: STRAIGHT UP

Step 10: Measure the force needed to move the cart straight up. No incline is needed for this task. Record the force in the data table.

Step 11: The distance the cart would travel along this path to move 20 cm above the tabletop is 20 cm. Record the distance (in meters) in the data table.

PART E: CALCULATING WORK

Step 12: Calculate the work done along the shallow incline. Multiply the force applied to the cart by the distance the cart traveled to move 20 cm above the tabletop. Show the calculation for the shallow path below, and transfer the result to your data table. Include the correct units in all your values.

$W = F \cdot d =$ _____ $\times$ _____ $=$ _____

Step 13: Calculate the work for the other paths: the medium incline, the steep incline, and straight up. Record the results in the data table.

Step 14: Show your instructor your completed data table before proceeding to the Summing Up section.

Data Table

	Angle θ (degrees)	Force F (newtons)	Distance d (meters)	Work W (joules)
Shallow Path				
Medium Path				
Steep Path				
Straight Up				

Summing Up

1. As the incline gets steeper, what happens to the force required to pull the cart?

   ___**The force increases significantly.**

   ___**The force decreases significantly.**

   ___**The force remains about the same.**

   (Compare the force needed to pull the cart along the shallow path with the force needed to pull the cart straight up. A significant difference is one that is 20% or greater.)

2. As the incline gets steeper, what happens to the distance traveled by the cart?

   ___**The distance increases significantly.**

   ___**The distance decreases significantly.**

   ___**The distance remains about the same.**

   (Compare the distance along the shallow path with the distance of the path straight up.)

3. As the incline gets steeper, what happens to the work required to move the cart 20 cm above the tabletop?

   ___**The work increases significantly.**

   ___**The work decreases significantly.**

   ___**The work remains about the same.**

   (Compare the work needed to move the cart up the shallow path with the work needed to move the cart straight up.)

4. What is the **advantage** of using an inclined plane rather than moving something straight up?

   _____

   _____

   _____

5. What is the **disadvantage** of using an inclined plane rather than moving something straight up?

   _____

   _____

   _____

6. The work done to move something is a measure of the energy required to complete the task. The energy required to move an automobile is provided by the fuel it consumes. Would it be more fuel-efficient to drive to the top of a hill along a steeply inclined road or a gradually inclined road? Explain your answer in terms of what you observed in this experiment.

   _____

   _____

   _____

CONCEPTUAL PHYSICAL SCIENCE | **Experiment**

Momentum and Energy | **Energy Conservation During Free Fall**

Dropping the Ball

Purpose
To determine and compare the potential energy of the ball before it's dropped and the kinetic energy of the ball after right before it hits the ground

Apparatus
acrylic tube (1.0 inch in diameter, about 4 feet in length)
small rare-earth magnet (neodymium or equivalent)
steel ball, about 16 mm
table
support rod
meterstick

2 photogates and timers
table clamp
2 three-finger clamps or buret clamps

SAFETY NOTE: Use caution when handling the magnet to avoid pinching. Keep it away from computers, sensitive electronic devices, and magnetic storage media such as computer disks.

Discussion
Your textbook gives several examples of work done in lifting an object, transforming the work to potential energy, and then transforming it to kinetic energy when the object falls. In this experiment you'll do the same with a steel ball. You'll measure its potential energy when it is lifted to a certain height. You'll measure its kinetic energy after it falls from that height. Then you'll compare the two energy values.

Procedure
Step 1: Arrange the apparatus as shown in Figures 1 and 2. The upper photogate (gate 1) should be about 5 cm above the lower photogate (gate 2). Connect both gates to the timer.

Step 2: Configure the photogate timer to read the time between the two gates. (That is, the timer starts when the beam of the first photogate is interrupted and stops when the beam of the second photogate is interrupted.)

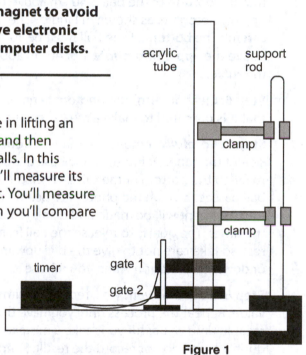

Figure 1

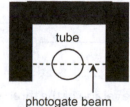

Figure 2. Top View
Make sure the photogate beam passes through the diameter of the tube.

DROPPING THE BALL

Step 3: Measure the distance between the photogate beams as shown in Figure 3. *Be very careful taking this measurement.* Record the distance in centimeters, and convert it to meters.

Distance between photogate beams:

$d =$ _____ cm

$=$ _____ m

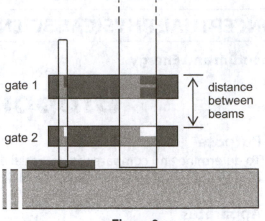

Figure 3

Step 4: Determine the mass of the ball. Record its mass in grams and kilograms.

Mass of steel ball:

$m =$ _____ g

$=$ _____ kg

Step 5: Set the ball inside the tube. Use the magnet to lift the ball up through the tube so that the bottom of the ball is 40 cm above the upper photogate, as shown in Figure 4. Be careful: The bottom of the ball must be 40 cm above the **upper photogate**, not 40 cm above the **table**.

Step 6: Clear or "arm" the photogate timer so that it is prepared to make a measurement.

Step 7: *Carefully* remove the magnet from the side of the tube. Doing so will release the ball to fall to the bottom of the tube. When the ball passes through the photogate beams, a measurement will be made and displayed on the timer. The goal is to release the ball from rest, so take care not to give the ball upward or downward motion when you release it.

Step 8: If the trial went well, record the time value. Repeat the process until you have three reliable time values. (If you make a mistake during a trial, do not record the result. Simply repeat the process until you have a good trial.)

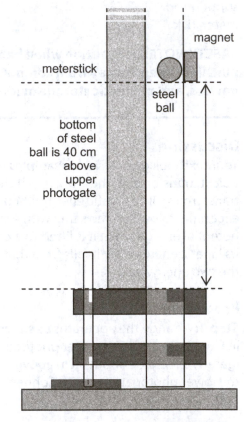

Figure 4

Time values: _____

_____

_____

Step 9: Determine the average of the three values, and record it in the data table.

Step 10: Repeat the process for drops from 60 cm, 80 cm, and 100 cm. Record the results in the data table.

Data Table

Drop Height h (meters)	Photogate Time t (seconds)	Speed v (m/s)	Potential Energy PE (J)	Kinetic Energy KE (J)
0	—	0	0	0
0.40				
0.60				
0.80				
1.00				

a. Calculate the potential energy of the ball when it was 40 cm above the photogate using the equation PE = mgh (m is the mass of the ball, g is 9.8 m/s^2, and h is 0.40 m). Show your work and your solution. (The value should be between 0.050 J and 0.100 J.) Record your solution in the data table as well.

_____

_____

_____

b. Calculate the potential energy of the ball at the other heights and record your solutions in the data table.

c. Calculate the speed of the ball after it fell 40 cm using the equation $v = \dfrac{d}{t}$ (d is the distance between the photogates and t is the time it took the ball to pass between the photogates). Show your work and your solution below. (The value should be between 2.50 m/s and 3.00 m/s.) Record your solution in the data table.

_____

_____

_____

d. Calculate the speed of the ball after falling from the other drop heights and record your solutions in the data table.

e. Calculate the kinetic energy of the ball after it fell 40 cm using the equation $KE = \frac{1}{2}mv^2$ (m is the mass of the ball, and v is the speed of the ball). Show your work and your solution. (The value should be between 0.050 and 0.100 J.) Record your solution in the data table.

_____

_____

_____

f. Calculate the kinetic energy of the ball after falling from the other drop heights, and record your solutions in the data table.

DROPPING THE BALL

Summing Up

1. When the drop height doubles from 40 cm to 80 cm, which of the following quantities also doubles (approximately)?
   ___ **speed after the fall**
   ___ **potential energy at the drop height**
   ___ **kinetic energy after the fall**

2. Which statement best describes the relationship between the potential energy at the drop height to the kinetic energy after the fall?
   ___ **The potential energy is always significantly higher than the kinetic energy.**
   ___ **The kinetic energy is always significantly higher than the potential energy.**
   ___ **The potential energy and kinetic energy are about the same.**
   (A significant difference in this experiment would be a difference of 20% or more. See the Percent Error Appendix for information on how to calculate percent difference.)

3. Use your findings to predict the following values for a trial involving dropping the ball from 160 cm.

   a. **Potential energy =** _____

   b. **Kinetic energy =** _____

   c. **Speed after falling =** _____

   Hint: By what factor was the "80-cm ball" faster than the "40-cm ball"? Use this factor to determine how much faster the "160-cm ball" will be compared with the "80-cm ball."

Going Further

When the ball was released from a particular height, its potential energy was transformed to kinetic energy as it fell. This type of energy transformation happens on a roller coaster as well. Potential energy that the roller coaster has at the top of the first hill is transformed to kinetic energy as it rolls downward. In an ideal roller coaster (with no frictional losses), the kinetic energy at the bottom of the hill would be equal to the potential energy at the top.

Consider an ideal roller coaster. The first hill has a certain height. When the roller coaster reaches the bottom of the hill, it is traveling at a certain speed.

1. How much higher would the hill have to be so that the roller coaster had twice as much **kinetic energy** at the bottom of the hill?

_____

2. How much higher would the hill have to be so that the roller coaster had twice as much **speed** at the bottom of the hill? (The answer to this question is not the same as the answer to the previous question.)

_____

3. In real-world roller coasters, each hill is shorter than the hill before it. Why do you suppose that is?

_____

_____

| **CONCEPTUAL PHYSICAL SCIENCE** | **Activity** |

Momentum and Energy   **Physics in the Soda Pop Geyser**

The Fountain of Fizz

Purpose
To observe a Mentos geyser and measure several quantities involved in the eruption, and then determine the speed of the fizz and the power of the eruption

Apparatus
2-liter bottle of diet soda pop
7 Mentos candies
mechanism for dropping the candies into the soda pop
electronic balance (or other means of measuring the duration of the eruption)
stopwatch
means of measuring the height of the eruption (which may exceed 5 meters)

Discussion
The Mentos geyser is a popular science demonstration. Several Mentos candies are dropped into an open bottle of diet soda pop, and the liquid rapidly turns to fizz. The fizz shoots out of the bottle to great heights. The chemistry involves the carbon dioxide in the soda pop. The nature of the surface of the Mentos candies and an ingredient (gum arabic) in the candy are the significant factors. They cause the carbon dioxide to come out of solution, creating the fizz. In this activity, you will focus on the *physics* of the demonstration.

Procedure
Step 1: Determine and record the mass of the demonstration ingredients *before* the demonstration. The ingredients are the 2-liter bottle of soda pop and the Mentos (seven Mentos seem to work well).

    Initial mass = _____

Step 2: Prepare the eruption in an open, *outdoor* space where splashing diet soda pop will not create a problem.

Step 3: Prepare a means to measure the maximum height of the eruption.

Step 4: Prepare to measure the duration of the eruption. That is, you will start timing when the fizz first emerges from the bottle and stop timing when the column of fizz collapses.

Step 5: When preparations are complete, activate the demonstration by dropping the candies into the soda pop, and observe the height and time of the eruption.

    Maximum height of eruption = _____

    Time of eruption = _____

Step 6: Measure and record the mass of the ingredients (Mentos, soda pop, and bottle) that remain after the eruption has concluded.

Final mass = _____

Summing Up

1. What type of energy (kinetic or potential) does the fizz have when it emerges from the bottle on its way up?

_____

2. What type of energy does the fizz have when it reaches the top of its flight and is about to come back down?

_____

3. How much mass was ejected from the bottle during the eruption?

_____

_____

4. Assume that all the mass ejected in the eruption rose to the maximum height measured during the demonstration. Calculate the potential energy of all the fizz at that height. (Start by writing the equation for potential energy.)

_____

5. According to the principle of conservation of energy, the potential energy of the fizz at the top of the flight is equal to the kinetic energy of the fizz when it emerges from the bottle. Write the equation for the kinetic energy of an object in terms of its mass and speed.

_____

6. Rearrange the equation to solve for the speed of the soda pop as it emerges from the bottle.

_____

_____

_____

7. Power is the rate at which work is done or energy is transformed. That is,

$$\text{Power} = \frac{\text{energy}}{\text{time}}$$

Use the energy found above and the time measured during the demonstration to calculate the power developed in the eruption.

_____

_____

_____

CONCEPTUAL PHYSICAL SCIENCE	Activity

Gravity, Projectiles, and Satellites

Connecting Mass to Weight

The Weight

Purpose
To investigate the relationship between weight and mass

Apparatus
spring scale (10-newton capacity)
slotted masses (one 100 g, two 200 g, one 500 g)
mass hanger
table clamp
support rod

rod clamp
crossbar (short rod)
collar hook
graph paper

Discussion
Mass and weight are different quantities. Mass is a measure of an object's inertia, the extent to which an object resists changes to its state of motion. Weight is a measure of the interaction between an object and the planet the object is nearest to; usually, that planet is Earth. The weight of an object is related to its mass. In this activity you will find out what that relationship is.

Procedure
Step 1: Check the calibration of the spring scale, and adjust it if necessary. The spring scale needs to read zero when there is no load on it. If you're not sure how to do this, ask your instructor how to calibrate the spring scale.

Step 2: Arrange the apparatus as shown in Figure 1.

Step 3: Determine the mass (in grams) and the weight of the mass hanger (in newtons—as shown on the spring scale). Record these values in the second row of the data table.

Step 4: Add 100 grams of slotted mass to the mass hanger. The total mass of the load on the spring scale is now 100 grams plus the mass of the mass hanger. Record the total mass and the weight value (shown on the spring scale) in the next row of the data table.

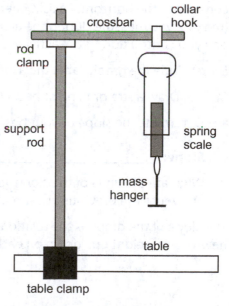

Figure 1

Step 5: Repeat the previous step with 200 g, 300 g, 400 g, 500 g, 600 g, 700 g, and 800 g of slotted mass. Remember that the *total* mass in each case is the sum of the slotted mass and the mass of the mass hanger. When you are finished, the column of mass values (in grams) and the corresponding column of weight values (in newtons) will be filled.

Data Table

Total Hanging Mass m (grams)	Total Hanging Mass m (kilograms)	Weight of Mass W (newtons)
0	0	0

Step 6: Convert the mass values to kilograms (1000 grams = 1 kilogram, so 237 grams = 0.237 kg, and so on).

Step 7: Make a plot of weight versus mass, using the mass values in kilograms. As is always the case, the first variable listed in the title of the graph constitutes the vertical axis; the second variable constitutes the horizontal axis. Generally speaking, the first variable listed is the dependent variable (the one you measure during the activity), and the second variable is the independent variable (the one you control during the activity).

Step 8: Title the graph; label the axes to correctly indicate the quantity, units, and scale of each axis.

Step 9: Draw a straight line of best fit through the data points plotted on your graph.

a.   Determine the slope of your best-fit line, and record it below.

   Slope: _____

b.   What are the units of the slope you found? The slope does have units! Show the units in your answer to the previous question.

The slope of the graph is the relationship between weight and mass. The slope tells how many newtons of weight pull down on each kilogram of mass.

Summing Up

1. What would have been the weight of 1.0 kg? Extend your best-fit line or use a ratio to determine the answer.

_____

_____

2. On the Moon, each kilogram of mass is pulled down with 1.6 newtons of weight. Add a dashed line to your graph showing the results that would have been obtained if this activity had been done on the Moon. How does the slope of the **Moon** line compare with the slope of the **Earth** line? Is it steeper (more nearly vertical) or shallower (more nearly horizontal)?

_____

_____

3. If the activity had been done on Jupiter, the resulting line would have had a steeper (more nearly vertical) slope. What does this tell you about how the strength of Jupiter's gravitational field compares with the strength of Earth's gravitational field?

_____

_____

_____

CONCEPTUAL PHYSICAL SCIENCE	Activity

Gravity, Projectiles, and Satellites   **Horizontal and Vertical Motion**

The Big BB Race

Purpose
To compare the path of a projectile launched horizontally with that of an object in free fall

Apparatus
simultaneous launcher/dropper
table clamp   support rod   right-angle clamp

Discussion
Suppose a ball bearing (BB) were launched horizontally at the same time that another BB was dropped from the same height. Which one would reach the ground first?

Procedure
Step 1: Draw the path you think the launched BB will take on Figure 1. That is, draw a line that connects the launch point and the impact point in the diagram and traces the path you think the BB will follow.

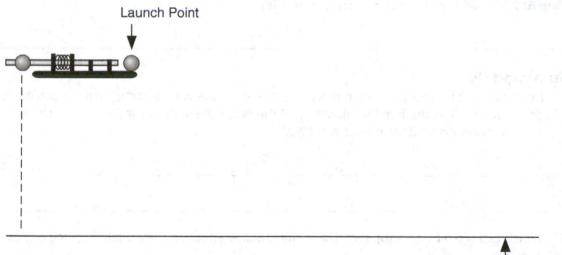

Figure 1

Step 2: Consider the following predictions of three students. Write arguments supporting each prediction (whether you agree with the prediction or not).

a. Student X predicts that the dropped BB will hit first. Why might Student X believe this?

_____

b. Student Y predicts that the launched BB will hit first. Why might Student Y believe this?

_____

c. Student Z predicts that the dropped BB and the launched BB will both hit at the same time. Why might Student Z believe this?

Step 3: One of the reasons sometimes offered to support the prediction that the dropped ball will hit first is that the launched ball will travel forward for some distance before starting to move downward. What factors might determine the length of this "no-fall distance," as shown in Figure 2?

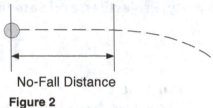

No-Fall Distance

Figure 2

Step 4: Which prediction do you agree with: dropped BB hits first, launched BB hits first, or both hit at the same time?

Step 5: Perform and observe the operation of the simultaneous launch/drop mechanism.

a. Observe a dropped BB.

b. Observe a launched BB.

c. Observe The Big BB Race—a simultaneous launch *and* drop.

Step 6: Which BB hits the ground first, or is it a tie?

Summing Up

1. How does the horizontal motion of a projectile affect the vertical motion of the projectile? In other words, does the horizontal motion of the projectile make it move faster or slower in the vertical direction (or does it have no effect)?

2. Which factors—if any—appear to have the greatest effect on the no-fall distance discussed above?

3. If the launched BB had had a rocket engine propelling it forward after it was launched, what—if anything—would have been different about the outcome of The Big BB Race?

CONCEPTUAL PHYSICAL SCIENCE	**Experiment**

Gravity, Projectiles, and Satellites   A Projectile Puzzle You <u>Can</u> Solve

Bull's Eye

Purpose
To predict the landing point of a projectile

Apparatus
1/2-inch (or larger) steel ball   table
empty can   meterstick
stopwatch
means of projecting the steel ball horizontally at a known velocity

Discussion

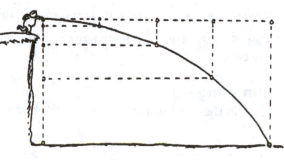

Figures 4.15–4.17 in *Conceptual Physical Science* show how projectiles move with constant speed in the horizontal direction while undergoing free-fall acceleration in the vertical direction. Figure 1 at the right shows the same for a ball tossed horizontally. The horizontal and vertical motions are ***independent*** of each other!

Figure 1

Instead of tossing a ball from a cliff, you'll fire a steel ball off the edge of a table and into a can on the floor below—all without a trial shot!

When engineers build bridges or skyscrapers, they do ***not*** do so by trial and error. For the sake of safety and economy, the effort must be right the ***first*** time. Your goal in this experiment is to predict where a steel ball will land when projected horizontally from the edge of a table. The final test of your measurements and computations will be to position an empty can so that the ball lands in the can on the ***first*** attempt.

Procedure

Step 1: Your instructor will provide you a means of projecting the steel ball at a known horizontal velocity (it may be a spring gun, a ramp, or some other device). Position the ball projector at the edge of a table so the ball will land downrange on the floor. Do ***not*** make any practice shots (the fun of this experiment is to ***predict*** where the ball will land without seeing a trial). Figure 2 shows the quantities involved in this experiment.

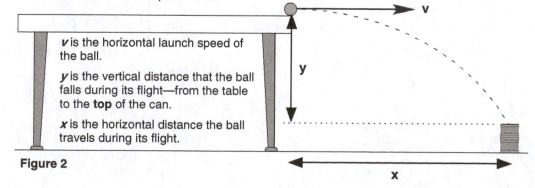

v is the horizontal launch speed of the ball.

y is the vertical distance that the ball falls during its flight—from the table to the **top** of the can.

x is the horizontal distance the ball travels during its flight.

Figure 2

You will be given the initial speed of the steel ball (or perhaps you'll have to devise a way to measure it). Record the firing speed here.

Horizontal speed $v =$ _____ **cm/s**

Step 2: Carefully measure the vertical distance y that the ball must drop from the bottom end of the ball projector in order to land in an empty soup can on the floor. Be sure to take the height of the can into account when you make this measurement.

Vertical distance $y =$ _____ **cm**

Step 3: The vertical free-fall motion is governed by the equation $y = (1/2)gt^2$. A rearrangement of the equation, $t = \sqrt{(2y/g)}$, enables you to calculate the time it takes the ball to fall from its original height. For gravitational acceleration, use $g = 980$ cm/s^2.

Time of flight $t =$ _____ **s**

Step 4: The range is the horizontal distance a projectile travels, x. Predict the range of the ball using $x = vt$. Write down your predicted range.

Predicted range $x =$ _____ **cm**

Now place the can on the floor where you predict it will catch the ball.

Step 5: Only after your instructor has checked your predicted range and your can placement, shoot the ball.

Summing Up

1. Did the ball land in the can on the first trial? If not, how many trials were required?

_____

2. What possible errors would account for the ball overshooting the target?

_____

3. What possible errors would account for the ball undershooting the target?

_____

4. If the can you used were replaced with a taller can, would you get a successful run if the ball started with a higher speed, a lower speed, or the same speed? Explain.

_____

_____

CONCEPTUAL PHYSICAL SCIENCE	Experiment
Gravity, Projectiles, and Satellites	Orbital Mechanics Simulation

Worlds of Wonder

Purpose
To use a simulation to study the orbital mechanics of a simplified solar system

Apparatus
computer
PhET simulation "My Solar System" (available at http://phet.colorado.edu)

Discussion
Lab activities involving stars and planets are difficult to conduct inside a classroom or laboratory. Because you cannot create stars and planets to experiment with in the classroom, you will use a computer simulation that uses the laws of gravity to show the behavior of large objects at great distances from one another.

Procedure
SETUP
Step 1: Turn on the computer and let it complete its start-up process.

Step 2: Open the PhET simulation "My Solar System." If you're not sure how to do this, ask your instructor for assistance.

Step 3: When the simulation opens, the screen should resemble the figure below.

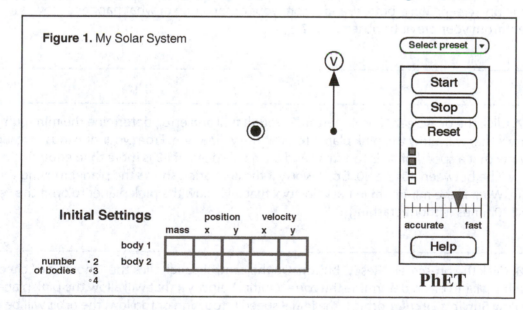

Figure 1. My Solar System

PART A: NEWTON'S CANNON
Isaac Newton explained that universal gravitation accounted for both the fall of an apple and the orbit of the Moon. At the time, this was hard for people to understand. Newton used a thought experiment to show how the same force could explain free-fall and orbital motion. In this activity, you will simulate "Newton's Cannon," illustrated in Figure 4.29 on page 108 of your text.

Step 1: In the control panel on the right side of the screen, the checkboxes for "System Centered" and "Show Tracks" should be checked. Set the "accurate/fast" slider to the midpoint. Set the "Initial Settings" for body 1 (yellow Sun) and body 2 (pink planet) as follows.

   a. **Body 1: mass = 200, position x = 0, position y = 0, velocity x = 0, velocity y = 0.**

   b. **Body 2: mass = 1, position x = 0, position y = 100, velocity x = 0, velocity y = 0.**

Step 2: Click the on-screen "Start" button, and record your observation.

_____

_____

Step 3: Click the on-screen "Reset" button to stop the simulation and restore the initial position and velocity settings.

Step 4: Change the initial Velocity x of body 2 (the pink planet) to 40. Click the on-screen "Start" button, and record your observation of what happens. How is it different from your previous observation?

_____

_____

Step 5: Click the on-screen "Reset" button to stop the simulation and restore the initial position and velocity settings. Change the initial Velocity x of the pink planet to 80. Click the on-screen "Start" button, and record your observation.

_____

_____

Step 6: Click the on-screen "Reset" button. Change the initial Velocity x of the pink planet to 160. Click the on-screen "Start" button, and record your observation of what happens. How is it different from your previous observation?

_____

_____

Step 7: Click the on-screen "Reset" button. Through trial and error, determine the minimum initial Velocity x that will allow the pink planet to orbit the yellow Sun. From your previous investigations, you know that a speed of 40 is too small and an initial speed of 80 is more than enough. So your result will be between 40 and 80. Don't worry if the animation shows the planet moving through the Sun. What is the minimum initial Velocity x that will allow the pink planet to orbit the yellow Sun at least 10 times without crashing?

_____

Step 8: Click the on-screen "Reset" button. On the control panel, click the "Show Grid" checkbox. Through trial and error, determine the correct initial Velocity x that will allow the pink planet to orbit the yellow Sun in a _circular_ orbit. If the initial speed is too high or too low, the orbit will be elliptical. What speed is just right to allow a _circular_ orbit?

_____

PART B: HARMONY OF THE WORLDS

There is a mathematical relationship between the orbital radius and the orbital speed of planets circling the Sun. German mathematician Johannes Kepler discovered this relationship. He started with volumes of astronomical data, worked through hundreds of pages of calculations, and spent approximately 30 years pursuing the discovery. In this activity, you'll use the simulation to generate data that will enable you to make the discovery in much less time.

Step 1: Find circular orbits for planets at various distances from the Sun. Start by setting the "position y" of the pink planet at a distance of 50. This sets the orbital radius to 50.

Step 2: On the control panel, click to activate the "Tape Measure."

Step 3: Click and drag the tape measure box icon until its crosshairs (+) are on the pink planet. Now click and drag the other end of the tape measure vertically downward, across the Sun, until it measures a distance of 100. Because you set the orbital radius to 50, the orbital diameter is 100. So the tape measure represents the diameter of the orbit.

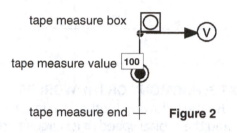

Step 4: Set the Velocity x of the pink planet to 150. Click the on-screen "Start" button and observe the orbit. Because the trace of the pink planet doesn't pass through the far end of the tape measure, the orbit is not circular.

Step 5: Click the on-screen "Reset" button. Try a different Velocity x for the pink planet. Through trial and error, keep trying until you find the speed that results in a circular orbit. The trace of the pink planet will pass through the far end of the tape measure when the orbit is circular. Record the Velocity x in the data table.

Step 6: Find circular orbits when the orbital radius is 100, 150, and 200 to complete the data table.

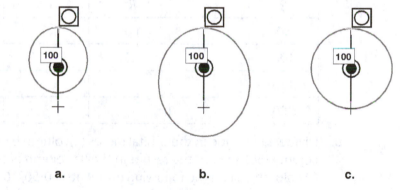

a.   b.   c.

Figure 3
a. and b. Noncircular Elliptical Orbits c. Circular Orbit

Data Table

Orbital Radius *R* (Position *y*)	Orbital Speed *v* (Velocity *x*)
50	
100	
150	
200	

Summing Up

PART A: NEWTON'S CANNON

1. A cannonball dropped from a cliff will fall straight down and hit Earth's surface. How could the cannonball be made to orbit Earth instead?

_____

_____

2. Based on your experience with the simulation, which do you think is more common: *circular* orbits or noncircular *elliptical* orbits? Defend your answer.

_____

_____

PART B: HARMONY OF THE WORLDS

3. Use the following method to determine the relationship between the orbital radius of a planet and the orbital speed of its circular orbit. For this activity, you'll limit your investigation to three possible relationships. They are as follows:

   Orbital radius is inversely proportional to orbital speed: $R \sim 1/v$.

   Orbital radius is inversely proportional to the square of orbital speed: $R \sim 1/v^2$.

   Orbital radius is inversely proportional to the square root of orbital speed: $R \sim 1/\sqrt{v}$.

   a. To see the pattern in the data, simplify and process your data. First, rewrite the orbital data in the table below.

R	v	R^*	v^*	$1/v^*$	$1/v^{*2}$	$1/\sqrt{v}^*$
50		1.00	1.00	1.00	1.00	1.00
100		2.00				
150						
200						

   b. Divide each value in the orbital radius (R)column by the first value in the orbital radius column (50). Record the results in the R* column of the table above. That is, the values in the R* column will be the following quotients: 50/50, 100/50, 150/50, and 200/50.

   c. Repeat this process using the orbital speed (v) data to determine values of v*. That is, divide all values of orbital speed by the first value of orbital speed.

   d. Now complete the last three columns by performing the appropriate mathematical operations on the values in the v* column.

4. Select the column that best matches the R* column. Is it ___$1/v^*$, ___$1/v^{*2}$, or ___$1/\sqrt{v}^*$?

5. Complete the statement: Orbital radius is inversely proportional to the

_____

> Johannes Kepler worked out the mathematics of orbits. Isaac Newton used Kepler's findings to develop the theory of universal gravitation!

CONCEPTUAL PHYSICAL SCIENCE	Activity

Fluid Mechanics   **Density and Flotation Simulation**

Pool Cubes: Density

Purpose
To rank items by mass, size, and density, and to determine a density-based rule for flotation

Apparatus
computer   PhET sim "Density" (available at http://phet.colorado.edu)

Discussion
Some pieces of gold are large and some are small. They have different masses and different sizes (volumes), but they all have the same density. And although a pound of lead and a pound of feathers have the same mass (on Earth), they have vastly different volumes because they have vastly different densities. The text defines density as the ratio of mass to volume:

$$\text{Density} = \frac{\text{mass}}{\text{volume}}$$

In this activity, you will explore the meaning of density using simulated solid cubes, an electronic balance, and a pool of water.

Procedure

Step 1: Launch the PhET sim "Density." In the on-screen Blocks panel, select "Mystery." You should see a screen similar to Figure 1.

Step 2: Use the on-screen electronic scale to determine the mass of the blocks. Record the masses in Table 1 below. Rank blocks by mass using the symbols of equality and inequality. For example, "A > B = C" would mean that A has a mass greater than that of B, and that B has a mass equal to that of C.

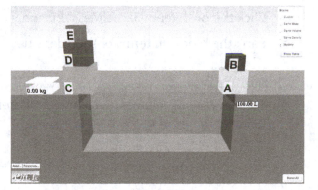

Figure 1. PhET Density Sim

Step 3:
a. Determine the volumes of the blocks, and record them in Table 1. Rank the blocks by volume (size) using the notation described above.

---

b. What was your technique for determining volume? "Eye-balling," or visual inspection, is not a sufficiently precise technique. Your technique must be valid for **all** blocks.

Table 1. Density Sim Data, Calculations, Observations, and Findings

Block	Mass (kg)	Volume (L)	Sink/Float?	Density	Material
A					
B					
C					
D					
E					

Step 4: Calculate the densities of the blocks. In the space below, show your calculation for the density of block A.

Step 5: Indicate whether each block will float or sink.

Step 6: Click the on-screen "Show Table" button, and record the name of the material for each block.

Step 7:

a. Click the on-screen "Same Volume" button.

b. What is the volume of each block? _____ L.

c. Rank the blocks in terms of their densities using the notation described above.

_____

Step 8:

a. Click the on-screen "Same Mass" button.

b. Rank the blocks in terms of their densities.

_____

c. The blue block behaves in a manner different from any other block in this activity. Describe the peculiar behavior of the blue block.

_____

_____

Summing Up

1. What condition—in terms of density—determines whether a block will sink or float?

_____

_____

2. Why does the blue block behave as it does?

_____

_____

CONCEPTUAL PHYSICAL SCIENCE   **Experiment**

Fluid Mechanics   **Buoyancy and Flotation Simulation**

Pool Cubes: Buoyancy

Purpose
To investigate the nature of the buoyant force and to see the role it plays in determining whether or not an object floats

Apparatus
computer   PhET sim "Buoyancy" (available at http://phet.colorado.edu)

Discussion
When objects are immersed in a fluid, the fluid exerts a force on them. This is the buoyant force.

Procedure
INITIAL SETUP
Turn the computer on, and let it complete its start-up cycle. Locate and open the PhET sim "Buoyancy," and note that it opens in its "Intro" tab. Do not change the settings in the on-screen Blocks, Fluids, or Show Forces panels.

PART A: WOOD
Step 1: Measure (in the sim) and record the weight (W) of the 5.00-kg wood block.

_____

Step 2:

a. Place the wood block in the water and notice that it floats.

b. When the wood block rests on the ground, the downward gravitational force is balanced by an upward normal ("contact") force. When the wood block is floating, the gravitational force is still there, but the normal ("contact") force is not. The force exerted by the fluid is the buoyant force (B). When the block is floating, which force on it is greater: the gravitational force or the buoyant force? Defend your answer.

_____

_____

_____

c. Record the value of the buoyant force acting on the wood block.

_____

Step 3:

a. What volume of water does the wood block displace when it floats in the water?

_____

b. What is the mass of that displaced water? (Note that the density of water is 1.00 kg/L.)

_____

_____

c. What is the weight of the displaced water?

_____

_____

PART B: BRICK

Step 1: Measure and record the weight (W) of the 5.00-kg block of brick (in air).

_____

Step 2:

a. Measure and record the **apparent weight** (N) of the brick when submerged in water.

_____

b. Does the brick appear to be   ___heavier or   ___lighter when submerged in water?

c. When submerged in water, the brick appears to have an additional force pushing it ___upward ___downward. This additional force is the **buoyant force** (B).

Step 3:

a. In the space to the right, draw and label a diagram of the brick block at rest on the bottom of the pool to show how the forces of weight, apparent weight, and buoyant force act on the block.

b. Calculate the size of the buoyant force. Show your work.

Step 4: What is the volume of the brick block? (Observe; do not calculate!)

_____

Step 5: What are the volume, mass, and weight of the displaced water?

Summing Up

1. How does the buoyant force acting on an object compare to the weight of the water displaced by that object?

_____

_____

2. The wood floated; the brick did not. What condition—in terms of forces—is required for an object to float?

_____

_____

3. What is the condition for floating in terms of object and fluid densities?

_____

_____

Going Further

1. **Fluid Density.** Replace the water with oil (using the on-screen Fluids panel). Describe and execute a method to determine the density of the oil.

_____

_____

_____

_____

2. **Latent Bloomer.** Switch the sim to the Buoyancy Playground tab. Select My Block from the on-screen control panel.

   a. What is the density of the densest block you can create?

   _____

   b. i. Place that block in the water. It __floats __sinks __is neutrally buoyant.

   ii. Now maximize its volume. It __floats __sinks __is neutrally buoyant.

   iii. While the block is at maximum volume, minimize its mass. It __floats __sinks __is neutrally buoyant.

   c. Without dragging the low-density block or changing its mass or volume, how can you bring it to rest at the **bottom** of the pool?

   _____

   _____

3. **Sim Surfing.** Click the on-screen Reset All button. Select Two from the on-screen Blocks panel. Set Block A to be an 8-L block of Styrofoam; set Block B to be a 2-L block of brick. Place the Styrofoam block in the water. Place the brick block on top of the Styrofoam block. The arrangement floats.

   a. What is the mass of the stack?

   _____

   b. What is the volume of the water displaced?

   _____

   c. What is the mass of the water displaced?

   _____

   d. Change the volume of the Styrofoam block to 6 L, 4 L, and 2 L. What do you observe during this process?

   _____

   _____

   e. Adjust the volume of the Styrofoam block until the stack has neutral buoyancy. Record the volume of the block.

   _____

   f. **"Above and Beyond" Challenge!** An object with a high density ρ_H can be floated on an object with a low density ρ_L in a fluid with a density ρ if the volumes of the objects (V_H and V_L) are right.

      i. The condition for such a stack having neutral buoyancy, in terms of masses (m_H and m_L) and volumes (V_H and V_L) of the blocks and the density ρ of the water, is expressed mathematically as

      $$\rho = (m_H + m_L) / (V_H + V_L)$$

      ii. What is mass m in terms of density ρ and volume V?

      iii. Eliminate m_H and m_L from the initial expression in part f.i. above. Solve to find the volume of the Styrofoam block V_L required so that if a brick block (whose volume V_H is known) is stacked on top of it, the stack will be neutrally buoyant in a fluid whose density is ρ. This is an algebraic solution to the puzzle you solved in the sim in part 3.d. Your solution may include ρ_H, V_H, ρ_L, V_L, ρ, and constants.

iv. Given $V_H = 2.0$ L, $\rho_H = 2.0$ kg/L, $\rho_L = 0.15$ kg/L, and $\rho = 1.0$ kg/L, what is V_L?

v. If the calculated value does not match the value found in the sim, ask your instructor for assistance.

g. i. Calculate the largest volume of aluminum that can be supported by a 10.0-L block of wood floating in gasoline. The density of aluminum is 2.70 kg/L, the density of wood is 0.40 kg/L, and the density of gasoline is 0.70 kg/L.

ii. Can you verify your solution using the sim?

CONCEPTUAL PHYSICAL SCIENCE

Activity

Fluid Mechanics

What Makes an Object Sink or Float?

Sink or Swim?

Purpose
To observe the effect of density on various objects placed in water

Apparatus
equally massive blocks of lead, wood, and Styrofoam
at least two aluminum cans of soda pop—both diet and regular
aquarium tank or sink two-thirds filled with water
egg
wide-mouth graduated cylinder
bowl
salt
spoon
balance

Discussion
Why can some people float in water while others can't? The answer has to do with density. This activity should increase your understanding of the role of density in floating.

Procedure
Step 1: Balance equal masses of lead and wood, using a double-beam balance. Repeat, using an equal mass of Styrofoam.

How do the **volumes** compare? How do the **densities** therefore compare?

_____

_____

Step 2: Try floating cans of soda pop in the aquarium/sink.

a. Which ones float? Which ones sink?

_____

_____

b. How does the density of the different kinds of soda pop compare with the density of tap water?

_____

_____

SINK OR SWIM?

c. Hypothesize how the relative densities are related to sugar content.

_____

_____

Step 3: Use a balance to measure the mass of an egg. Using a wide-mouth graduated cylinder, carefully determine the volume of the egg by measuring the volume of water it displaces when it is slowly (gently!) lowered into the graduated cylinder. Calculate its density: $d = m/V$.

Density = _____

Step 4: Now try to float the egg in a bowl of water. Does it float? If not, dissolve salt in the water until the egg floats.

a. How does the density of an egg compare with that of tap water?

_____

_____

b. With that of salt water?

_____

_____

Summing Up

1. Does adding salt to the water make the water less dense or more dense? How?

_____

_____

_____

2. Why do some people find it difficult to float while others don't? What evidence can you cite for the notion that it is easier to float in salt water than in fresh water?

_____

_____

_____

_____

CONCEPTUAL PHYSICAL SCIENCE

| **Activity** |

Fluid Mechanics

Flotation

Boat Float

Purpose
To investigate Archimedes' principle and the principle of flotation

Apparatus
spring scale
string and masking tape
600-mL beaker
clear container or 3-gallon bucket
chunk of wood
toy boat capable of carrying a 1200-gram
  cargo, or 9-inch aluminum cake pan

100-g mass
balance
rock or hook mass
500-mL graduated cylinder
water
modeling clay
3 lead masses or lead fishing weights

Discussion
An object submerged in water takes up space and pushes
water out of the way—the water is **displaced.**
Interestingly enough, the water that is pushed out of the
way pushes back on the submerged object. For example,
if the object pushes a volume of water with a
weight of 100 N out of its way, then the water reacts by pushing back on the object with a force of
100 N—Newton's third law. The object is **buoyed** upward with a force of 100 N. This is summed up
in Archimedes' principle, which states that the **buoyant force** that acts on any completely or
partially submerged object **is equal to the weight of the fluid the object displaces.**

Procedure
Step 1: Use a spring scale to determine the weight of an object (rock or hook mass) that is first out
of water and then under water. The difference in weights is the buoyant force. Record.

Weight of object out of water = _____

Weight of object in water = _____

Buoyant force on object = _____

Step 2: Devise a method to find the volume of water displaced by the object. Record the volume of
water displaced. Compute the mass and weight of this water. (Remember, 1 mL of water has a mass
of 1 g and a weight of 0.01 N.)

Volume of water displaced = _____

Mass of water displaced = _____

Weight of water displaced = _____

BOAT FLOAT

How does the buoyant force on the submerged object compare with the weight of the water displaced?

_____

_____

Note: To simplify calculations for the remainder of this activity, measure and determine **masses,** without the need of finding their equivalent **weights** ($W = mg$). Keep in mind, however, that an object floats because of a buoyant **force.** This force is due to the **weight** of the water displaced.

Step 3: Use a balance to measure the mass of a piece of wood, and record the mass in the data table. Measure the volume of water displaced when the wood floats. Record the volume and mass of water displaced in the data table.

a. How are the buoyant force on any floating object and the weight of the object related?

_____

_____

b. How does the mass of the wood compare with the mass of the water displaced?

_____

_____

c. How does the buoyant force on the wood compare with the weight of the water displaced?

_____

_____

Step 4: Add a 100-g mass to the wood so that the wood displaces more water but still floats. Measure the volume of water displaced, and calculate its mass. Record them in the data table.

How does the buoyant force on the wood with its 100-g load compare with the weight of the water displaced?

_____

_____

Step 5: Roll the clay into a ball and find its mass. Measure the volume of water it displaces after it sinks to the bottom of a graduated cylinder. Calculate the mass of water displaced. Record all volumes and masses in the data table.

a. How does the mass of water displaced by the clay compare with the mass of the clay out of the water?

_____

_____

b. Is the buoyant force on the submerged clay greater than, equal to, or less than its weight out of the water? What is your evidence?

_____

_____

Step 6: Retrieve the clay from the bottom, and mold it into a shape that enables it to float. Sketch or describe this shape. Measure the volume of water displaced by the floating clay. Calculate the mass of the water, and record that value in the data table.

Data Table

Object	Mass (g)	Volume of Water Displaced (mL)	Mass of Water Displaced (g)
Wood			
Wood + 50-g Mass			
Clay Ball			
Floating Clay			

Summing Up

1. Does the clay displace more, less, or the same amount of water when it floats as it did when it sank?

_____

2. Is the buoyant force on the floating clay greater than, equal to, or less than the weight of the clay?

_____

3. What can you conclude about the weight of an object and the weight of water displaced by the object when it floats?

_____

4. Is the buoyant force on the clay ball greater when it is submerged near the bottom of the container or when it is submerged near the surface? What is your evidence?

_____

_____

5. Is the pressure that the water exerts on the clay ball greater near the bottom of the container than when the clay ball is submerged near the surface? Exaggerate the depth involved and cite expected evidence.

_____

_____

6. Why are your last two answers different?

_____

BOAT FLOAT

Going Further

1. Suppose you are on a ship in a canal lock. If you throw a ton of lead bricks overboard from the ship into the canal lock, will the water level in the canal lock go up, go down, or stay the same? Write down your prediction **before** you proceed to Step 8.

    Prediction for water level in canal lock: _____

2. Float a toy boat loaded with lead "cargo" in a relatively deep container filled with water (deeper than the height of the lead masses). For observable results, the container should be just slightly bigger than the boat. Mark and label the water levels on masking tape placed on the container and on the sides of the boat. Remove the masses from the boat and put them in the water. Mark and label the new water levels.

3. What happens to the water level **on the side of the boat** when you remove the cargo? What does this say about the amount of cargo carried by ships that float high in the water?

    _____

4. What happens to the water level **in the container** when you place the cargo in the water? Explain why this happens.

    _____

    _____

5. Similarly, what happens to the water level in the canal lock when the bricks are thrown overboard?

    _____

6. Suppose the freighter is carrying a cargo of Styrofoam instead of bricks. What happens to the water level in the canal lock if the Styrofoam (which floats in water) is thrown overboard?

    _____

7. When a ship is launched at a shipyard, what happens to the sea level all over the world—no matter how imperceptibly?

    _____

8. When a ship in the harbor launches a little rowboat from the dock, what happens to the sea level all over the world—no matter how imperceptibly?

    _____

9. Cite evidence to support your (different?) answers to Questions 7 and 8.

    _____

10. One of the most fascinating applications of a floating object displacing a weight of water equal to its own weight is the Falkirk Wheel, which is featured in the photo opener to Chapter 5 and Figure 5.17 in your text. Explain why only a low-power motor can lift heavy ships in this way.

    _____

    _____

    _____

CONCEPTUAL PHYSICAL SCIENCE	Experiment

Thermal Energy and Thermodynamics   **Kinetic Theory Simulation**

Bouncing off the Walls

Purpose
To control and observe the behavior of gas particles (atoms or molecules) as modeled in a simulation to investigate properties of gas such as temperature and pressure

Apparatus
computer   PhET simulation "Gas Properties" (available at http://phet.colorado.edu)

Discussion
Kinetic molecular theory explains the large-scale characteristics of gases in terms of the behavior of the atoms and molecules that make up the gas. The "Gas Properties" simulation lets you see the individual particles in motion. It gives you control of a chamber of gas and shows you the effects of the changes you make.

Setup
Step 1: Start the computer and log in. Open the PhET simulation "Gas Properties."

Step 2: In the on-screen control panel, click the "Measurement Tools" button.

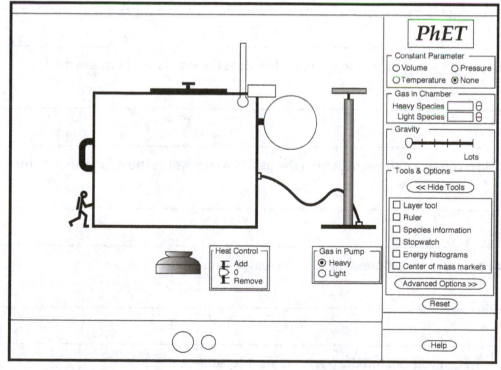

Step 3: Add labels to the figure above for each item listed below.

Thermometer	Pressure gauge	Heat source/sink	Gas pump
Chamber lid	**Play/pause**	**"Scubie" (the volume adjuster)**	

Procedure

PART A: SIMULATION MECHANICS (*Remember that any time you need to, you can use the on-screen "Reset" button to return to the initial setup.*)

Step 1: Determine two distinct methods by which you can add particles to the chamber.

a. Method 1: Use the pump. How do you manipulate the pump handle to get the **greatest** number of particles into the chamber in **one** stroke?

_____

_____

b. Describe Method 2: How can you **precisely** control the number of particles injected into the chamber? (Method 2 does not involve direct use of the pump.)

_____

_____

Step 2: How can you release particles from the chamber (**without** breaking the chamber)?

a. Method 1:

_____

_____

b. Method 2 (**completely** different from Method 1):

_____

_____

Step 3: How can you add heat to the gas? How does the simulation illustrate this?

_____

_____

Step 4: How can you remove heat from the gas? How does the simulation illustrate this?

_____

_____

Step 5: How can you compress the gas (decrease its volume)?

_____

_____

Step 6: How can you expand the gas (increase its volume)?

_____

_____

PART B: THE NATURE OF THE IDEAL GAS LAW
Step 1: What happens to the pressure in the chamber when you heat the gas?

_____

Step 2: What happens to the temperature in the chamber when you compress the gas?

_____

Step 3: Locate the "Constant Parameter" section of the on-screen control panel. Lock the temperature. Use "Scubie" to slowly compress the gas. What happens to the temperature, and what action is taken (by the simulation) to maintain constant temperature?

_____

_____

Step 4: Lock the pressure. (Doing so releases the lock on temperature.) Add heat. What happens to the pressure, and what action is taken (by the simulation) to maintain constant pressure?

_____

_____

PART C: ALL SPECIES GREAT AND SMALL
Create a chamber that contains a mix of 50 light and 50 heavy gas particles. In the on-screen control panel, activate "Species information."

Step 1: Consider the species information and the meaning of temperature.

a.   Which species—if either—has the greater average speed?

_____

b.   Does the temperature in the chamber reflect the average speed, the momentum, the potential energy, or the kinetic energy of the particles? And how does this explain your finding in Step 1.a.?

_____

_____

Step 2: Reset the chamber to have about 100 heavy particles. The temperature should be 300 K. Click the on-screen button to activate "Center of mass markers."

a.   Consider the following statement: "The average **_speed_** of air molecules in a room may be over 1000 mph (400 m/s), while their average **_velocity_** is approximately zero." The particles in the simulation's chamber are modeling the air molecules in a room. Cite evidence from the simulation to confirm or reject the quoted statement.

_____

_____

_____

b. What is the name of the condition that prevails when the average velocity of atmospheric molecules around you is not zero? (*Hint:* It's a common four-letter word starting with the letter "w.")

_____

Going Further

1. Under what specific condition will the lid be blown off the chamber? Is it possible to blow the lid off with just one particle in the chamber?

_____

_____

_____

2. Is it possible to achieve a six-figure temperature? If so, how? If not, what's the highest temperature you could achieve?

_____

_____

_____

3. Low-temperature physicists have not yet been able to cool anything to a temperature of absolute zero (0 K).
   a. Cool a simulated 50/50 (50 light and 50 heavy particles) gas sample to absolute zero.
   b. When the temperature hits absolute zero, are the particles shown to be at rest?

_____

_____

4. Use Scubie and the ice ("Heat Control: Remove") to compress and cool a 50/50 sample to the smallest volume possible and to absolute zero. Now use Scubie to rapidly expand the volume of the chamber all the way out. Describe what happens. And which species wins the race to the far side of the chamber (where Scubie is)?

_____

_____

_____

_____

_____

CONCEPTUAL PHYSICAL SCIENCE **Experiment**

Thermal Energy and Thermodynamics **Water's Heat Capacity**

Temperature Mix

Purpose
To predict the final temperature of a mixture of cups of water at different temperatures

Apparatus
3 Styrofoam cups
liter container with a wide mouth
thermometer (Celsius)
pail of cold water
pail of hot water

Discussion
If you mix a pail of cold water with a pail of hot water, the final temperature of the mixture will be between the two initial temperatures. What information would you need to predict the final temperature? You'll begin with the simplest case of mixing **equal** masses of hot and cold water.

Procedure
Step 1: Begin by marking your three Styrofoam cups equally at about the three-quarter mark. You can do this by pouring water from one container to the next and marking the levels along the inside of each cup.

Step 2: Fill the first cup to the mark with hot water from the pail, and fill the second cup with cold water to the same level. Measure and record the temperature of both cups of water.

Temperature of cold water = _____

Temperature of warm water = _____

Step 3: Predict the temperature of the water when the two cups are combined. Then pour the two cups of water into the liter container, stir the mixture slightly, and record its temperature.

Predicted temperature = _____

Actual temperature of water = _____

If there was a difference between your prediction and your observation, what may have caused it?

Pour the mixture into the sink or waste pail. (Don't be a klutz and pour it back into either of the pails of cold or hot water!) Now you'll investigate what happens when **unequal** amounts of hot and cold water are combined.

Step 4: Fill one cup to its mark with cold water from the pail. Fill the other two cups to their marks with hot water from the other pail. Measure and record their temperatures. Predict the temperature of the water when the three cups are combined. Then pour the three cups of water into the liter container, stir the mixture slightly, and record its temperature.

      Predicted temperature = _____

      Actual temperature of water = _____

Pour the mixture into the sink or waste pail. Again, do **not** pour it back into either of the pails of cold or hot water!

a. How did your observation compare with your prediction?

_____

b. Which of the water samples (cold or hot) changed more when it became part of the mixture? In terms of energy conservation, suggest a reason why this happened.

_____

Step 5: Fill two cups to their marks with cold water from the pail. Fill the third cup to its mark with hot water from the other pail. Measure and record their temperatures. Predict the temperature of the water when the three cups are combined. Then pour the three cups of water into the liter container, stir the mixture slightly, and record its temperature.

      Predicted temperature = _____

      Actual temperature of water = _____

Pour the mixture into the sink or waste pail. (At this point, you and your lab partners may simply pour waste water back into either of the pails of cold or hot water.)

a. How did your observation compare with your prediction?

_____

b. Which of the water samples (cold or hot) changed more when it became part of the mixture? Suggest a reason why this happened.

_____

_____

Summing Up

1. What determines whether the equilibrium temperature of a mixture of two amounts of water will be closer to the temperature of the initially cooler water or to that of the initially warmer water?

_____

2. How does the formula $Q = cm\Delta T$ apply here?

_____

CONCEPTUAL PHYSICAL SCIENCE	**Experiment**

Thermal Energy and Thermodynamics   **Comparing Specific Heat Values**

Spiked Water

Purpose
To determine which is better able to increase the temperature of a quantity of water: a mass of hot nails or the same mass of equally hot water

Apparatus
equal-arm balance (Harvard trip balance or equivalent)
4 large insulated cups
bundle of short, stubby nails tied together with string
thermometer (Celsius)
hot and cold water
paper towels

Discussion
Suppose you have cold feet when you go to bed, and you want something to keep your feet warm throughout the night. Would you prefer to have a bottle filled with hot water or one filled with an equal mass of nails at the same temperature as the water? The one that can store more thermal energy will do the better job. But which one is it? In this experiment, you'll find out.

Procedure
Consider the equal masses of nails and water, both warmed to the same temperature. Which one will do a better job of heating a sample of cold water? Make a prediction before performing the experiment.

Step 1: Fill two cups 1/3 full of **cold** water. Place one cup on each pan of the balance to make sure they contain equal masses of water, as shown in Figure 1. Add water to the lighter side until they do. Make sure your bundle of nails can be completely submerged in the cold water.

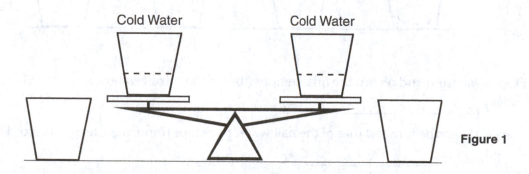

Cold Water   Cold Water

Figure 1

Step 2: Dry the nails.

Step 3: Place a large empty cup on each pan of a beam balance. Place the bundle of nails into one of the cups. Add **hot** water to the other cup until it balances the cup of nails, as shown in Figure 2. When the two cups are balanced, the masses in the two cups are the same.

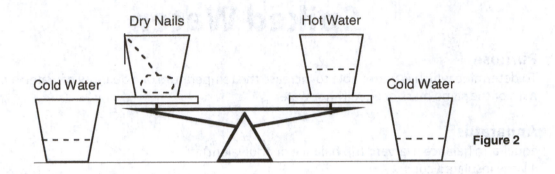

Figure 2

Step 4: Lower the bundle of nails into the hot water as shown in Figure 3. Be sure that the nails are completely submerged in the hot water. Allow the nails and the water to reach thermal equilibrium. (This will take a minute or two.)

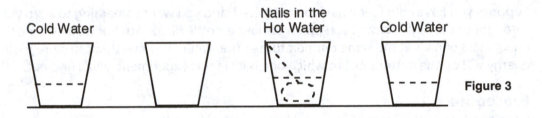

Figure 3

Step 5: Once the nails and the hot water have come to thermal equilibrium, take the nails out of the hot water and put them in one of the cups of cold water. Pour the remaining hot water into the other cup of cold water. See Figure 4.

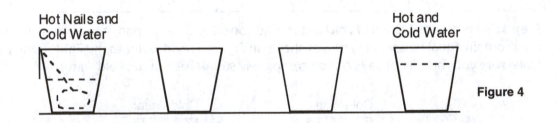

Figure 4

Step 6: Measure and record the final temperature of the mixed water.

$T_W =$ _____ °C

Step 7: When the temperature of the nail-water mix stops rising, measure and record it.

$T_N =$ _____ °C

Summing Up

1. Which was hotter before being put into the cold water, the nails or the hot water in which the nails were soaking? How do you know?

   _____

   _____

2. Which was more effective in raising the temperature of the cold water, the hot nails or the hot water?

   _____

   _____

3. Suppose you have cold feet when you go to bed, and you want something to warm your feet throughout the night. Would you prefer to have a bottle filled with hot water or one filled with an equal mass of nails at the same temperature as the water? Relate your answer to your findings in this experiment.

   _____

   _____

4. A student who conducted this experiment suggested that the temperature of the nail-water mix rose more than it otherwise would have because some water clung to the nails when they were transferred from the hot water to the cold water. Another student says that this caused the temperature of the nail-water mix to rise less than it would have if no water had clung to the nails during the transfer. Which student do you believe and why?

   _____

   _____

   _____

Name _____ Section _____ Date _____

CONCEPTUAL PHYSICAL SCIENCE

Heat Transfer and Change of Phase   **Thermal Absorption**

Canned Heat: Heating Up

Purpose
To compare the ability of different surfaces to absorb thermal radiation

Apparatus
heat lamp and base   radiation cans: silver, black, and white
thermometer   access to cold tap water
paper towel   graph paper

Discussion
Does the color of a surface make a difference in how well it absorbs thermal radiation? If so, how? The answer to these questions could help you decide what to wear on a hot, sunny day and what color to paint your house if you live in a hot, sunny climate. In this experiment, you will compare the thermal absorption ability of three surfaces: silver, black, and white. You'll do this by filling cans that have these surfaces with water and then exposing the cans to heat lamps. You'll measure the temperature of the water in each of the cans while they're being exposed to the heat and see whether there's a difference in the rate at which the temperatures increase.

Pre-lab Questions
1. In which of the three cans do you think the water will heat up at the fastest rate?

_____

2. In which of the three cans do you think the water will heat up at the slowest rate?

_____

Procedure
Step 1: Arrange the apparatus so that the heat lamp will shine equally on all three cans. The cans should be about 1 foot in front of the lamp. Do not turn the light on yet.

Step 2: Fill the cans with cold water and wipe up any spills. Quickly measure the initial temperature of the water in each of the cans and record the numbers in the data table.

Step 3: Turn on the lamp and start timing. Place the thermometer in the silver can.

Step 4: At the 1-minute mark (1:00), read the temperature of the water in the silver can, and record it in the data table. Quickly move the thermometer to the black can. Gently swirl the water in the can with the thermometer.

Step 5: At the 2-minute mark (2:00), read the temperature of the water in the black can, and record it in the data table. Quickly move the thermometer to the white can. Gently swirl the water in the can with the thermometer.

Step 6: At the 3-minute mark (3:00), read the temperature of the water in the white can, and record it in the data table. Quickly move the thermometer to the silver can. Gently swirl the water in the can with the thermometer.

Step 7: Repeat Steps 4–6 until the 21-minute temperature reading is made.

Data Table

Silver Can Temperatures (°C)	Black Can Temperatures (°C)	White Can Temperatures (°C)
Initial	Initial	Initial
T at 1:00	T at 2:00	T at 3:00
T at 4:00	T at 5:00	T at 6:00
T at 7:00	T at 8:00	T at 9:00
T at 10:00	T at 11:00	T at 12:00
T at 13:00	T at 14:00	T at 15:00
T at 16:00	T at 17:00	T at 18:00
T at 19:00	T at 20:00	T at 21:00

Step 8: Turn off the heat lamp. Plot your data for all three cans on a single temperature vs. time graph.

Step 9: Determine the change in temperature for the water in each can while the heat lamp was on.

a. Determine the temperature change in the silver can while the heat lamp was on. Subtract the 1-minute temperature reading from the 19-minute temperature reading.

 Temperature change in silver can: _____°C

b. Determine the temperature change in the black can while the heat lamp was on. Subtract the 2-minute temperature reading from the 20-minute temperature reading.

 Temperature change in black can: _____°C

c. Determine the temperature change in the white can while the heat lamp was on. Subtract the 3-minute temperature reading from the 21-minute temperature reading.

 Temperature change in white can: _____°C

Summing Up

1. In which can did the water heat up at the fastest rate? In which can did the water heat up at the slowest rate? Did your observations match your predictions?

 _____

2. Which would be a better choice if you were going to spend a long time outdoors on a hot, sunny day: a black T-shirt or a white T-shirt?

 _____

3. What happens to the thermal radiation that falls on each of the cans? Is it absorbed or reflected?

 Silver: _____ **Black:** _____ **White:** _____

CONCEPTUAL PHYSICAL SCIENCE | **Experiment**

Heat Transfer and Change of Phase | Thermal Emission

Canned Heat: Cooling Down

Purpose
To compare the ability of different surfaces to radiate thermal energy

Apparatus
radiation cans: silver and black
thermometer
graph paper

paper towel
access to hot water

Discussion
Does the color of a surface make a difference in how well it radiates thermal energy? If so, how? The answer to these questions could help you decide what color coffeepot will best keep its heat and what color might be used to radiate heat away from a computer chip. In this experiment, you will compare the thermal radiation ability of two surfaces: silver and black. You'll do this by filling two cans with these surfaces with hot water and then allowing them to cool down. You'll measure the temperature of the water in both cans while they're cooling down and see whether there's a difference in the rate at which the temperatures decrease.

Pre-lab Questions
1. In which of the two cans do you think the water will cool down at the faster rate?

_____

2. In which of the two cans do you think the water will cool down at the slower rate?

_____

Procedure
Step 1: Carefully fill the cans with hot water and wipe up any spills. Quickly measure the initial temperature of the water in each of the cans, and record it in the data table. Place the thermometer in the silver can.

Step 2: At the 1-minute mark, read the temperature of the water in the silver can, and record it in the data table. Quickly move the thermometer to the black can. Gently swirl the water in the can with the thermometer.

Step 3: At the 2-minute mark, read the temperature of the water in the black can, and record it in the data table. Quickly move the thermometer to the silver can. Gently swirl the water in the can with the thermometer.

Step 4: Repeat Steps 2 and 3 until the 20-minute temperature reading is made.

Data Table

Silver Can Temperatures (°C)	Black Can Temperatures (°C)
Initial	Initial
T at 1:00	T at 2:00
T at 3:00	T at 4:00
T at 5:00	T at 6:00
T at 7:00	T at 8:00
T at 9:00	T at 10:00
T at 11:00	T at 12:00
T at 13:00	T at 14:00
T at 15:00	T at 16:00
T at 17:00	T at 18:00
T at 19:00	T at 20:00

Step 5: Plot your data for both cans on a single temperature vs. time graph.

Step 6: Determine the change in temperature for the water in each can while it was allowed to cool.

a. Determine the temperature change in the silver can. Subtract the 1-minute temperature reading from the 19-minute temperature reading.

Temperature change in silver can: _____°C

b. Determine the temperature change in the black can. Subtract the 2-minute temperature reading from the 20-minute temperature reading.

Temperature change in black can: _____°C

Summing Up

1. In which can did the water cool down at the faster rate? In which can did the water cool down at the slower rate? Did your observations match your predictions?

_____

2. What color should the surface of a coffeepot be in order to keep the hot water inside it hot for the longest time?

_____

3. The central processing units (CPUs) in personal computers sometimes get very hot. To prevent damage and improve processing speed, they need to be kept cool. Often, a piece of metal is placed in contact with the CPU to draw some heat away. The metal then radiates its heat to the surroundings. To best remove heat from the CPU, should the metal's surface be colored black or left silvery?

_____

CONCEPTUAL PHYSICAL SCIENCE	Activity

Heat Transfer and Change of Phase   **Atmospheric Pressure and Boiling Point**

Cooling by Boiling

Purpose
To see that water will boil when pressure is lowered

Apparatus
400-mL beaker   hot plate (or equivalent to heat water above 60°C)
thermometer   vacuum pump with bell jar

Discussion
Whereas evaporation is a change of phase from liquid to gas at the surface of a liquid, boiling is a rapid change of phase at and below the surface of a liquid. The temperature at which water boils depends on atmospheric pressure. Have you ever noticed that water reaches its boiling point in a **shorter** time when you are camping up in the mountains? And have you noticed that at high altitude it takes **longer** to cook potatoes or other food in boiling water? That's because water boils at a lower temperature when the pressure of the atmosphere on its surface is reduced. Let's see!

Procedure
Step 1: Turn on the hot plate, and warm 200 mL of water in a 400-mL beaker to a temperature above 60°C. Record the temperature. Then place the beaker under and within the bell jar of a vacuum pump. If a thermometer will fit underneath the bell jar, place a thermometer in the beaker. Turn on the pump. What happens to the water?

    T = _____

_____

_____

Was the water **really** boiling?

_____

_____

Step 2: Stop the pump and remove the bell jar. What is the temperature of the water now?

    T = _____

Step 3: As time permits, repeat the procedure, starting with other temperatures, such as 80°C, 40°C, and 20°C, and record the time it takes for boiling to begin.

Summing Up

1. In terms of energy transfer, what does it mean to say that boiling is a cooling process? What cools?

_____

_____

_____

_____

2. Give two ways to cause water to boil.

_____

_____

3. On a hot stove, boiling water remains at a constant 100°C temperature. How is this observation evidence that boiling is a cooling process?

_____

_____

_____

_____

4. In Figure 7.33 in the text, water is seen to freeze by rapid evaporation. How is that observation related to this demonstration activity?

_____

_____

_____

CONCEPTUAL PHYSICAL SCIENCE	Experiment

Heat Transfer and Change of Phase   **Heat of Fusion**

Warming by Freezing

Purpose
To measure the heat released when freezing occurs

Apparatus
heat-generating pouch (supersaturated sodium acetate pack)
hot plate
large pan or pot of boiling water

Discussion
To liquefy a solid or vaporize a liquid, it is necessary to add heat. In the reverse process, heat is released when a gas condenses or a liquid freezes. Thermal energy that accompanies these changes of state is called *latent heat of vaporization* (going from gas to liquid or liquid to gas) and *latent heat of fusion* (going from liquid to solid or solid to liquid).

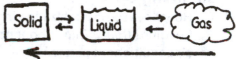

Energy is absorbed when change of phase is in this direction

Energy is released when change of phase is in this direction

Water, which normally freezes at 0°C (32°F), can be found under certain conditions in a liquid state at temperatures as low as –40°C (–40°F) or lower. This **supercooled** water (liquid water below 0°C) often exists as tiny cloud droplets, which are common in clouds where snow or ice particles form. Freezing in clouds depends on the presence of *ice-forming nuclei,* most of which are active in the –10°C to –20°C range. Ice-forming nuclei comprise many different substances, such as dust, bacteria, other ice particles, and the silver iodide used to "seed" clouds during droughts. Silver iodide is active at temperatures as high as –4°C.

Cold clouds containing large amounts of supercooled water and relatively small amounts of ice particles can be dangerous to aircraft. The skin (outer layer) of the aircraft, well below freezing, provides an excellent surface on which supercooled water suddenly freezes. This so-called aircraft icing can be quite severe under certain conditions.

The heat pack provides a dramatic example of a supercooled liquid. What you observe in the heat pack is actually the release of the latent heat of crystallization, which is analogous to the release of latent heat of vaporization or the latent heat of fusion. The freezing temperature of the sodium acetate solution inside the heat pack is about 55°C (130°F), yet it exists at room temperature. The heat pack can be cooled down to temperatures as low as –10°C before it finally freezes.

It only takes a quick click to activate the heat pouch. You will notice that the internal trigger button has two distinct sides. If you use your thumb and forefinger and squeeze quickly, you will not need to worry about which side is up.

After observing the crystallization of the sodium acetate and the heat released, you might want to try it again and measure the heat of crystallization. The package has a mass of about 146 grams. The packaging and the trigger mechanism have a mass of about 26 grams. Thus, the sodium acetate solution inside the package has a mass of about 120 grams.

Procedure

Step 1: Place the heat pack in an insulated container (such as a larger Styrofoam cup) with 400 g of room-temperature water. Allow the water and heat pack to reach equilibrium.

Step 2: Measure and record the water's initial temperature.

T_{low} = _____

Step 3: Activate the heat pack button. After 10 or 15 seconds, maximum temperature will be obtained.

Step 4: Return the heat pack to the container and measure the temperature of the water at regular intervals. Record the maximum temperature attained by the water.

T_{high} = _____

Step 5: How much heat is gained by the water?

Q = _____

Summing Up

1. How does the crystallization inside the heat pack relate to heating and air conditioning in a building? (*Hints:* Think about how steam heat and radiators operate or how the refrigerants in an air conditioner operate.)

_____

_____

_____

2. Speculate about how these processes relate to airplane safety?

_____

_____

_____

3. Think of and list some practical applications of the heat pouch.

_____

_____

_____

_____

> Wait a minute — before this we cooled by boiling, and now we warm by freezing? Must be some interesting physics going on here!

CONCEPTUAL PHYSICAL SCIENCE	Activity

Heat Transfer and Change of Phase   **Conduction and Absorption**

I'm Melting, I'm Melting

Purpose
To observe the curious heat transfer abilities of different surfaces

Apparatus
defrosting tray
a second defrosting tray wrapped tightly in aluminum foil
white Styrofoam plate
black Styrofoam plate (or a white plate blackened completely using a felt-tip pen)
4 ice cubes of similar size
paper towel

Discussion
If you walk around inside your house with bare feet, you probably notice that a tile floor feels much colder than a carpeted floor or rug. It's hard to believe that they might actually have the same temperature. The tile feels colder because it is a better conductor than carpet. Heat is conducted from your warmer feet to the cooler floor faster when the floor is tile than when the floor is carpet. Therefore, your feet are cooled faster by tile than they are by carpet at the same cool temperature. In this activity, you will see which kinds of surfaces transfer heat most rapidly.

Procedure
Step 1: Set the defrosting tray, defrosting tray covered in aluminum foil, white Styrofoam plate, and blackened Styrofoam plate on your table.

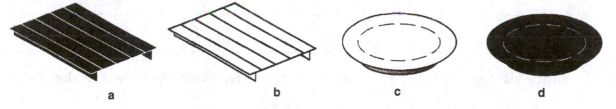

a   b   c   d

Figure 1. a. defrosting tray, **b.** defrosting tray covered in aluminum foil,
c. white Styrofoam plate, **d.** blackened Styrofoam plate

Which surfaces feel colder and which ones feel warmer? (Just touch a corner; don't transfer too much of your own body heat to any of the objects.)

Step 2: In a moment, you'll set an ice cube on each of the surfaces. When you do, the cubes will begin melting. Before you set the ice cubes on the surfaces, make some predictions.
a.   Which ice cube will melt most quickly?

b. Which ice cube will melt most slowly?

_____

Step 3: Set the ice cubes out on their respective surfaces quickly. Observe the ice cubes for several minutes (preferably until the fastest-melting ice cube melts completely).

a. Which ice cube melted most quickly?

_____

b. Which ice cube is melting most slowly?

_____

Summing Up

1. How do your observations compare with your predictions?

_____

_____

2. Which way did heat flow in this activity? (That is, from what did heat flow to what?)

_____

3. Two of the surfaces were conductors, and two of the surfaces were insulators. Which were which?

_____

_____

4. What advantage—in terms of heat transfer—did one defrosting tray have over the other?

_____

_____

5. What advantage—in terms of heat transfer—did one Styrofoam plate have over the other?

_____

_____

6. Which of the following conclusions are supported by your observations and which are not? Give evidence from this activity to justify your conclusion.

    a. "Metals transfer heat faster than Styrofoam."

_____

    b. "Black surfaces transfer heat faster than nonblack surfaces."

_____

CONCEPTUAL PHYSICAL SCIENCE   Experiment

Static and Current Electricity   **Conductors, Insulators, and Charge Transfer**

Electroscopia

Purpose

To use an electroscope to investigate electric charge polarity, electrostatic induction, and the difference between conductors and insulators

Apparatus

electroscope (can-form, *or* flask-form, *or* gold leaf)
2 acetate strips (transparent plastic)
silk cloth square (thin and soft)
small metal tube or rod

electrophorus
2 vinyl strips (opaque plastic)
wool cloth square (thick and coarse)
hair dryer (optional)

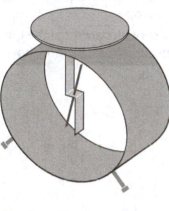

Can-form Electroscope

Flask-form Electroscope

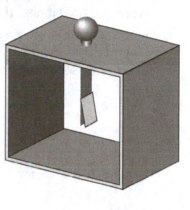

Gold Leaf Electroscope

Discussion

A microscope enables you to see things that are very small, and a telescope enables you to see things that are very far away. An electroscope enables you to detect the presence of excess electric charge. Your lab probably has **one** of the three types of electroscopes shown above. All three perform the same function. The nature of electric charge, charge transfer, and conductors and insulators can be investigated through the use of the electroscope.

Procedure

PART A: POLARITY

Step 1: Check to ensure that the electroscope and the charging materials are functioning correctly.

Figure 1. Neutral Electroscope
Can-form, flask-form, and gold leaf

a. When the electroscope is neutral and no charged objects are nearby, its pointer or foil leaf(s) should be "undeflected." See the simplified illustrations in Figure 1.

Note: You'll have only one type of electroscope. When you make your own illustrations later, use the diagram that corresponds to your electroscope type.

b. When a vinyl strip is vigorously rubbed with wool cloth and brought nearby, the pointer or leaf(s) should deflect. See the simplified illustrations in Figure 2. The deflected electroscope indicator shows the presence of excess electric charge.

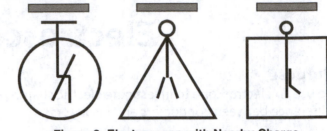

Figure 2. Electroscope with Nearby Charge
Can-form, flask-form, and gold leaf

c. When an acetate strip is rubbed vigorously with silk cloth and brought nearby, the electroscope indicator should again deflect.

Step 2: Vinyl + Vinyl

a. Rub a strip of vinyl with wool and hold it near the top of the electroscope. Keep it there.

b. While the first vinyl strip remains near the top of the electroscope, rub a second vinyl strip with wool and hold it near the first strip, close to the top of the electroscope.

c. In the space below, use words and illustrations to record your observations. (Show the type of electroscope you are using, and sketch two illustrations—before and after—to show how adding the second strip affects the electroscope.)

Step 3: Vinyl + Acetate

a. Rub a strip of vinyl with wool and hold it near the top of the electroscope. Keep it there.

b. While the vinyl strip remains near the top of the electroscope, rub an acetate strip with silk and hold it near the vinyl strip, close to the top of the electroscope.

c. In the space below, use words and illustrations to record your observations. (Sketch before and after illustrations to show how adding the second strip affects the electroscope.)

Step 4: Acetate + Acetate

a. Rub a strip of acetate with silk and hold it near the top of the electroscope. Keep it there.

b. While the first acetate strip remains near the top of the electroscope, rub a second acetate strip with silk and hold it near the first strip, close to the top of the electroscope.

c. In the space below, use words and illustrations to record your observations. (Show the type of electroscope you are using, and sketch before and after illustrations.)

Step 5: Acetate + Vinyl

a. Rub a strip of acetate with silk and hold it near the top of the electroscope. Keep it there.

b. While the acetate strip remains near the top of the electroscope, rub a vinyl strip with wool and hold it near the acetate strip, close to the top of the electroscope.

c. In the space below, use words and illustrations to record your observations. (Show the type of electroscope you are using, and sketch before and after illustrations.)

Step 6: Interpret these observations.

a. When one vinyl charge is added to another vinyl charge, do they tend to enhance one another or to cancel one another? Answer and describe your electroscope-based evidence.

_____

_____

b. When one acetate charge is added to another acetate charge, do they tend to enhance one another or to cancel one another? Answer and describe your electroscope-based evidence.

_____

_____

c. When vinyl and acetate charges are added to one another, do they tend to enhance one another or to cancel one another? Answer and describe your electroscope-based evidence.

_____

_____

> **Scientists refer to two types of charge. For historical reasons, the charge on silk-rubbed acetate is called *positive*, and the charge on wool-rubbed vinyl is called *negative*.**

PART B: DETERMINING THE SIGN OF AN UNKNOWN CHARGE

Step 1: Based on what you learned in Part A, describe a method for determining the type (sign) of the charge on a charged object. You know the sign of the charge on wool-rubbed vinyl, and you know the sign of the charge on silk-rubbed acetate. But how can you determine sign of the charge on any object using what you've learned so far?

_____

_____

_____

_____

Step 2: Charge the electrophorus. See Figure 3. *Pay very close attention to the details described in each step, and follow the sequence exactly as described.*

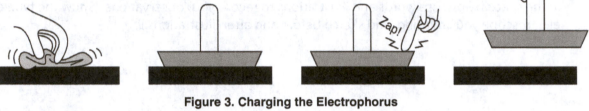

Figure 3. Charging the Electrophorus

a. Rub the electro-phorus **base** or a plastic chair seat with the appropriate cloth (silk on acrylic; wool on vinyl or Styrofoam).

b. Set the electro-phorus **plate** on the base or chair seat. Avoid metal rivets in the seats.

c. Touch the top of the metal plate. You may feel a slight shock; you may feel nothing. **Do not skip this step!**

d. Pick up the plate holding only the plastic handle. If the plate discharges, repeat the previous steps.

Step 3: Carry out the procedure you described in Step 1 to determine the charge on your electrophorus. What is the charge on the electrophorus, and what is your evidence?

_____

_____

_____

Step 4: Verify that your finding is correct. If you used an acrylic (transparent plastic) base, then your electrophorus plate carried a negative charge. If you used a vinyl (opaque plastic) or Styrofoam base, then your electrophorus plate carried a positive charge.

Step 5: Ask your instructor to bring an "unknown charged *object*" near your electroscope for investigation.

a. Describe the object.

_____

b.   What is the sign of the charge on the object, and what is your evidence?

_____

_____

_____

Step 6: Describe a method for determining the type (sign) of charge on a charged electroscope. In other words, if your electroscope indicator were deflected with no charged objects nearby, how could you determine the sign of the excess charge on the electroscope?

_____

_____

_____

_____

Step 7: Neutralize your electroscope by touching the top plate or metal ball. Ask your instructor to transfer an "unknown *charge*" onto your electroscope.
a.   Describe the method by which your instructor charged your electroscope.

_____

_____

b.   What was the sign of the charge transferred to your electroscope, and what is your evidence?

_____

_____

_____

PART C: ELECTROSTATIC INDUCTION
Step 1: Neutralize your electroscope so that the indicator is undeflected.

Step 2: Recall and record the sign of the charge on wool-rubbed vinyl: _____.

Step 3: Rub a strip of vinyl with wool, and bring it near to the top plate or ball of the electroscope. Observe the corresponding deflection of the indicator.

Step 4: While keeping the vinyl in place near the electroscope, *briefly* touch the top plate or ball of the electroscope. Observe that the indicator reverts to its undeflected state.

Step 5: Now move the vinyl away from the electroscope. What do you observe?

_____

_____

Step 6: Illustrate the entire sequence in the space below.

a. The electroscope is neutral; the indicator is undeflected.

b. Wool-rubbed vinyl is brought nearby.

c. With the vinyl in place, the top plate or ball of the electroscope is briefly touched.

d. The vinyl is removed.

Step 7: Using your procedure from Step 6 of Part B: Determining the Sign of an Unknown Charge, determine the sign of the charge left on the electroscope.

a. What is the sign of the charge left on the electroscope? ___positive ___negative

b. How does this compare to the sign of the charge on the vinyl? ___the same ___opposite

> **Charging an object without direct contact is called *induction*.**

PART D: CONDUCTORS AND INSULATORS

Conductors and insulators differ in several ways. Plastics, such as vinyl and acetate, are insulators. Metals, such as iron and aluminum, are conductors.

Step 1: Select the best descriptions of conductors and insulators, based on what you've observed so far.

___**Conductors can be charged.**   ___**Insulators can be charged.**

___**Conductors are always charged.**   ___**Insulators are always charged.**

___**Conductors are never charged.**   ___**Insulators are never charged.**

Step 2: Neutralize the electroscope by touching the top metal plate or ball. Charge a strip of plastic (vinyl or acetate—whichever works better for you). Place the charged plastic strip in direct contact with the top metal plate or ball of the electroscope. Then move the strip away from the electroscope and set it aside. In the space below, describe what happened using words and illustrations.

Step 3: Neutralize the electroscope by touching the top metal plate or ball. Charge the electrophorus (using the acrylic base or chair seat—whichever works better for you). Place the charged metal plate in direct contact with the top metal plate or ball of the electroscope. Then move the electrophorus plate away from the electroscope and set it aside. In the space below, describe what happened using words and illustrations.

Step 4: Describe the important difference in behavior between the charged insulator (plastic strip) and the charged conductor (metal plate) in the previous steps.

_____

_____

_____

Charging an object by direct contact is called *conduction*.

Summing Up

1. An object is brought near the top metal plate or ball of a neutral electroscope. The indicator deflects. From this, we may conclude that (select all that apply)
    ___the object carries a positive charge.   ___the object carries a negative charge.
    ___the object carries a charge, but the sign of the charge is unknown.
    ___the object is a conductor.   ___the object is an insulator.
    ___the object may be either a conductor **or** an insulator.

2. An object is brought near the top metal plate or ball of a neutral electroscope. The indicator deflects. The object touches the top of the electroscope and is then removed. The indicator on the electroscope remains deflected. From this, we may conclude that (select all that apply)
    ___the object carries a positive charge.   ___the object carries a negative charge.
    ___the object carries a charge, but the sign of the charge is unknown.
    ___the object is a conductor.   ___the object is an insulator.
    ___the object may be either a conductor **or** an insulator.

3. A strip of charged acetate is held near a neutral electroscope, causing the indicator to deflect. An object with an unknown charge approaches the top of the electroscope, and the indicator's deflection *increases*. From this, we may conclude that (select all that apply)
    ___the object carries a positive charge.   ___the object carries a negative charge.
    ___the object carries a charge, but the sign of the charge is unknown.
    ___the object is a conductor.   ___the object is an insulator.
    ___the object may be either a conductor **or** an insulator.

4. An object is brought near the top metal plate or ball of a neutral electroscope. The indicator deflects. The metal plate or ball at the top of the electroscope is briefly touched. The object is then removed. The indicator on the electroscope remains deflected. From this, we may conclude that (select all that apply)
    ___the object carries a positive charge.   ___the object carries a negative charge.
    ___the object carries a charge, but the sign of the charge is unknown.
    ___the object is a conductor.   ___the object is an insulator.
    ___the object may be either a conductor **or** an insulator.

5. A strip of charged vinyl is held near a neutral electroscope, causing the indicator to deflect. An object with an unknown charge approaches the top of the electroscope, and the indicator's deflection *decreases*. From this, we may conclude that (select all that apply)
    ___the object carries a positive charge.   ___the object carries a negative charge.
    ___the object carries a charge, but the sign of the charge is unknown.
    ___the object is a conductor.   ___the object is an insulator.
    ___the object may be either a conductor **or** an insulator.

6. An object is brought near the top metal plate or ball of a neutral electroscope. The indicator deflects. The object is then removed, touched, and returned. The indicator on the electroscope again deflects. From this, we may conclude that (select all that apply)

___the object carries a positive charge.   ___the object carries a negative charge.

___the object carries a charge, but the sign of the charge is unknown.

___the object is a conductor.   ___the object is an insulator.

___the object may be either a conductor **or** an insulator.

7. Based on your observations in this experiment, what is the difference between the behavior of conductors and that of insulators?

_____

_____

8. **Charging the Electrophorus**

The electrophorus is charged by induction. Show the charges in the electrophorus plate in Figure 4 to fully illustrate the movement of charge during this process.

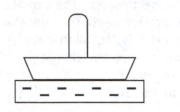 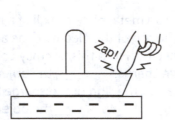 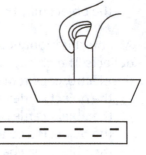

Figure 4. Charging the Electrophorus by Induction

a. When the electrophorus plate is placed on a negatively charged base, charge within the plate separates. The plate becomes polarized. (Positive charge is attracted to the base; negative charge is repelled from it.)

b. When the electrophorus is touched, negative charge escapes to you and on to Earth itself; positive charge stays with the electrophorus.

c. The electrophorus is left with a charge opposite to that of the charged base.

9. Suppose that charging an electrophorus left the plate with a positive charge.
   a. How could you use the positively charged electrophorus to place a positive charge on the electroscope?

_____

_____

   b. How could you use the positively charged electrophorus to place a negative charge on the electroscope?

_____

_____

_____

CONCEPTUAL PHYSICAL SCIENCE | **Activity**

Static and Current Electricity | The Van de Graaff Generator

Charging Ahead

Purpose
To observe the effects and behavior of static electricity

Apparatus
2 new balloons
Van de Graaff generator
several pie tins (small: 5-in diameter preferred)
several Styrofoam bowls
bubble-making materials (see note to right)

Going Further
wooden matches

> Bubble solution and a wand are OK, but a bubble gun is easier to work with and more fun!

Discussion
Scuff your feet across a rug and reach for a doorknob, and zap—electric shock! The electric charge that makes up the spark can be several thousand volts, which is why technicians have to be so careful when working with tiny circuits such as those in computer chips!

Procedure
Step 1: Blow up a balloon. After stroking it against your hair, place it near some small pieces of Styrofoam or puffed rice. Then place the balloon against the wall where it will "stick," as shown on the right. On the drawing, sketch the arrangement of some sample charges on the balloon and on the wall.

Step 2: Blow up a second balloon. Rub both balloons against your hair. Do they attract or repel each other?

_____

_____

Step 3: Stack several pie tins on the dome of the Van de Graaff generator. Turn the generator on. What happens and why?

_____

_____

_____

_____

Step 4: Turn the generator off, and discharge it with the discharge ball or by touching it with your knuckle. Stack several Styrofoam bowls in the generator, and turn the generator on. What happens and why?

_____

_____

_____

Step 5: With the Van de Graaff generator off and discharged, blow some bubbles toward it. Observe the behavior of the bubbles. Then turn the generator on and blow bubbles toward it again. Watch carefully. What happens and why?

_____

_____

_____

Step 6: Stand on an isolation stand (or rubber mat) next to a discharged Van de Graaff generator. Place one hand on the conducting sphere on top of the generator, and have your partner switch on the generator motor. Shake your head as the generator charges up. What do you experience?

_____

_____

_____

Summing Up

Which of the demonstrations in this activity are better explained by the principle that like charges repel and opposites attract? Which are better explained in terms of the differences between conductors and insulators?

_____

_____

_____

Going Further

Light a wooden match and move it near a charged generator. What happens and why?

_____

_____

CONCEPTUAL PHYSICAL SCIENCE	**Experiment**

Static and Current Electricity   **Connect Meters and Determine Resistance**

Ohm, Ohm on the Range

Purpose

To arrange a simple circuit, attach a meters to the circuit, measure current and voltage in the circuit, and find the relationship among current, voltage, and resistance

Apparatus

variable DC power supply (0–6 V)
2 power resistors with different resistances (values between 3 Ω and 10 Ω recommended)
power resistor with unknown resistance (for the "Going Further" section of the experiment)
miniature light bulb in socket (14.4-V flashlight bulb recommended)
DC ammeter (0–1 A analog recommended)
DC voltmeter (0–10 V analog recommended)
5 connecting wires
graph paper

Discussion

The current, voltage, and resistance in an electric circuit have a very specific relationship to one another. Designers of electric circuits must take this relationship into account, or their circuits will fail. This relationship is as important and fundamental in electricity as Newton's second law of motion is in mechanics. In this experiment, you will determine this relationship.

Procedure

PART A: CONNECTING THE METERS

Step 1: With the power supply turned down to zero, arrange a simple circuit using the power supply, the miniature bulb, and two connecting wires as shown in Figure 1. If your power supply has "AC" terminals, do not use them; connect only to the "DC" terminals.

Step 2: Verify that the circuit is working properly by slowly turning the knob on the power supply to increase the power to the circuit. The bulb should begin to glow and increase in brightness as power is increased. If the circuit doesn't work, make adjustments so that it does. Ask your instructor for assistance if necessary.

Step 3: Turn the power knob down to zero. (If there is a separate "on/off" switch, or if the knob can be clicked off, there is no need to power off. Simply turn the level down to zero.)

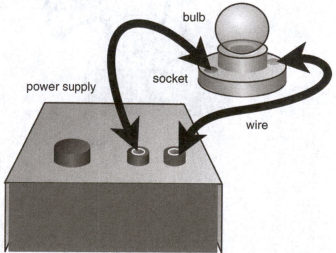

Figure 1. The Simple Circuit

Step 4: Connect the ammeter in series as shown in Figure 2. If the meter has multiple positive (+) terminals, use the one labeled "1." Slowly increase the power to the circuit, and observe the ammeter and the bulb. If the needle on your ammeter ever moves in the wrong direction (that is, tries to "go negative"), reverse the connections **on the meter** and try again.

When the connections are correct, what do you observe? Discuss the light—if any—from the bulb and the movement—if any—of the needle on the meter. You may get light with no meter activity, meter activity with no light, or both light and meter activity. If you get **no** light and **no** meter activity, ask your instructor for assistance.

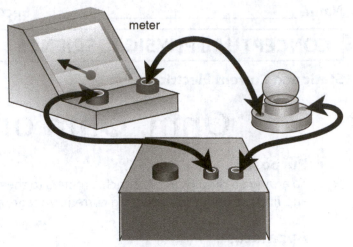

Figure 2. Meter Connected in Series

_____

_____

_____

Step 5: Turn the power back down to zero. Disconnect the ammeter, and restore the circuit to its original configuration (shown in Figure 1).

Step 6: Connect the ammeter in parallel as shown in Figure 3. Slowly increase the power to the circuit, and observe the ammeter and the bulb.

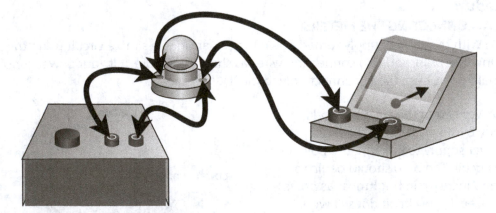

Figure 3. Meter Connected in Parallel

What do you observe? Discuss the light—if any—from the bulb and the movement—if any—of the needle on the meter. (Remember to switch the connections to the meter if the needle appears to move backward.)

_____

_____

_____

Step 7: Turn the power back down to zero. Disconnect the ammeter, and restore the circuit to its original configuration (shown in Figure 1).

Step 8: Connect the voltmeter in series as shown in Figure 2. Slowly increase the power to the circuit, and observe the voltmeter and the bulb. If the needle on your voltmeter ever moves in the wrong direction (tries to "go negative"), reverse the connections and try again.

When the connections are correct, what do you observe?

_____

_____

_____

Step 9: Turn the power back down to zero. Disconnect the voltmeter, and restore the circuit to its original configuration (shown in Figure 1).

Step 10: Connect the voltmeter in parallel as shown in Figure 3. Slowly increase the power to the circuit, and observe the voltmeter and the bulb.

What do you observe?

_____

_____

_____

Step 11: Synthesize your findings about connecting the meters.

a. When the ammeter is connected correctly, the circuit behaves as it did when no meters were connected (increasing the power increases the brightness of the bulb). When the power is increased, the ammeter shows increased current in the circuit. When the ammeter is connected incorrectly, the bulb remains dim or does not light at all, even though the ammeter shows significant current.

   The correct method is to connect the ammeter to the circuit in ___series.   ___parallel.

b. When the voltmeter is connected correctly, the circuit behaves as it did when no meters were connected (increasing the power increases the brightness of the bulb). When the power is increased, the voltmeter shows increased voltage in the circuit. When the voltmeter is connected incorrectly, the bulb remains dim or does not light at all, even though the voltmeter shows significant voltage.

   The correct method is to connect the voltmeter to the circuit in ___series.   ___parallel.

Step 12: Connect a circuit that includes the bulb, the ammeter, and the voltmeter. Use no more than five wires. Connect the ammeter and voltmeter correctly, based on your findings. Make no more than two connections to the power supply. Sketch a diagram in the space below to show how the wires connect the various circuit elements. Do not cross the lines representing wires in your sketch. Label the power supply, bulb, ammeter, and voltmeter.

Step 13: Verify that the circuit is working correctly. When the power is increased, the bulb should get brighter and **both** meter readings should increase. Show your instructor your working circuit.

PART B: COLLECTING DATA

Step 1: Turn the power down to zero. Replace the bulb with one of the known resistors. (Remove the bulb and socket from the circuit; connect the resistor in its place.)

Step 2: Record the Ω-value of the known resistor in the heading of the column for the first known resistor in the data table below. The Ω-value is written on the resistor and is between 3 and 10.

Step 3: Increase the power until the current indicated on the ammeter is 0.10 A.

Step 4: Record the corresponding voltmeter reading in the appropriate space in the data table.

Step 5: Repeat Steps 3 and 4 for current values of 0.20 A through 0.60 A (**or the highest value you can get on the ammeter if you can't get all the way up to 0.60 A**).

Step 6: Turn the power down to zero. Replace the first resistor with the second resistor.

Step 7: Repeat Steps 2–5 to complete the data table for the second resistor.

Data Table

Current I (amperes)	1st Known Resistor: _____ Ω Voltage V (volts)	2nd Known Resistor: _____ Ω Voltage V (volts)	
0	0	0	
0.10			
0.20			
0.30			
0.40			
0.50			
0.60			

Step 8: On your graph paper, plot graphs of voltage vs. current for both resistors. Plot both data sets on one set of axes. Voltage will be the vertical axis; current will be the horizontal axis. Scale the graph to accommodate all your data points. Label each axis as to its quantity and units of measurement.

Step 9: Make a line of best fit for each data set plotted on the graph.

Step 10: Determine the slope of each best-fit line.

What is the slope of each best-fit line? Identify each slope by its corresponding resistor value. Don't forget to include the correct units for each slope value.

_____–Ω **resistor best-fit line slope =** _____

_____–Ω **resistor best-fit line slope =** _____

Observe the similarity between the values of the slope and the values of the corresponding resistance.

Going Further

Obtain a power resistor with an unknown resistance from your instructor. Using the techniques of this experiment, determine the resistance of the resistor. Record your data and calculations in the space below.

Once you determine the resistance of the unknown resistor, ask your instructor for the accepted value. Record that value, and calculate the percent error in your value.

Summing Up

1. Which mathematical expression shows the correct relationship among current, voltage, and resistance?

___ $R = IV$ ___ $R = I/V$ ___ $R = V/I$

The correct answer is one form of the equation known as Ohm's law.

2. Suppose voltage vs. current data were taken for two devices, A and B, and the results were plotted to form the best-fit lines shown in Figure 4. Which device has the greater resistance? How do you know?

_____

_____

_____

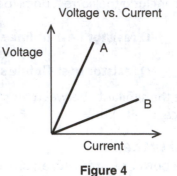

Voltage vs. Current

Figure 4

3. Suppose the voltage vs. current data for electrical device C produced graph C shown in Figure 5. How does the resistance of device C change as the current increases? Explain.

_____

_____

_____

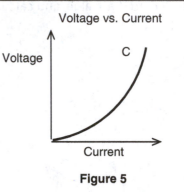

Voltage vs. Current

Figure 5

4. Suppose the voltage vs. current data for electrical device D produced graph D shown in Figure 6. How does the resistance of device D change as the current increases? Explain.

_____

_____

_____

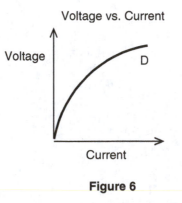

Voltage vs. Current

Figure 6

CONCEPTUAL PHYSICAL SCIENCE	**Activity**

Batteries and Bulbs

Purpose
To explore various arrangements of batteries and bulbs, as well as the effects of those arrangements on bulb brightness

Apparatus
2 C-cell or D-cell batteries
4 connecting wires
2 miniature bulbs (screw-base flashlight bulbs, 3-V to 6-V)
2 miniature bulb sockets

Discussion
Many devices include electronic circuitry, much of it quite complicated. Complex circuits are made, however, from simple circuits. In this activity, you will build one of the simplest yet most useful circuits ever invented—that for lighting a light bulb!

Procedure
Step 1: Remove the bulb from the mini-socket.

a. In the space below, draw a detailed diagram of the bulb, showing the following parts of the bulb's "anatomy."

    glass bulb   *filament*   *screw base*   *base contact (made of lead)*
    lead separator (glass bead)   *base-contact insulation ring*
    filament leads (tiny wires that lead to the filament)

b. There are four parts of the bulb's anatomy that you can touch (without having to break the bulb). Two of them are made of **conducting** material (metal), and two are made of **insulating** material. List them.

Conducting parts on the outside of the bulb: _____

Insulating parts on the outside of the bulb: _____

Step 2: Examine the two diagrams of a working electric circuit shown below. The diagram on the left shows pictorial representations of circuit elements. The diagram on the right shows symbolic representations. Use the symbolic representations in the steps that follow.

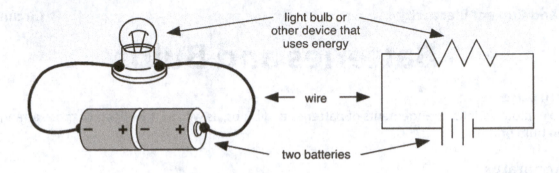

Using a bare bulb (out of its socket), one battery, and **two** wires, try lighting the bulb in as many ways as you can. On a separate sheet, sketch at least two **different** arrangements that work. Also sketch at least two arrangements that don't work. Be sure to label each of them either **works** or **doesn't work.**

Step 3: Using a bare bulb (out of its socket), one battery, and **one** wire, try lighting the bulb in as many ways as you can.

a. Sketch your arrangements, and note the ones that work.

b. Is it possible to light the bulb using the battery and **no** wires? Explain.

Step 4: Connect one bulb (in its socket) to two batteries as shown in Figure 1. This arrangement is often referred to as a **simple** circuit.

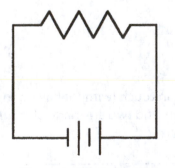

Figure 1. Simple Circuit

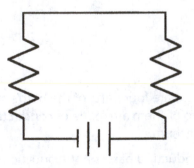

Figure 2. Series Circuit

Step 5: Connect the bulbs, batteries, and wire as shown in Figure 2. When the bulbs are connected one after the other like this, the result is a **series** circuit.

a. How does the brightness of each bulb in the series circuit compare with the brightness of the bulb in a simple circuit?

_____

b. What happens if one of the bulbs in a series circuit is removed? (Do this by unscrewing a bulb from its socket.)

_____

Step 6: Connect the bulbs, batteries, and wire as shown in Figure 3. When the bulbs are connected along separate paths like this, the result is a **parallel** circuit.

a. How does the brightness of each bulb in the parallel circuit compare with the brightness of the bulb in a simple circuit?

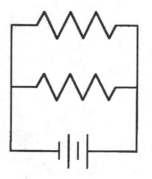

Figure 3. Parallel Circuit

_____

_____

b. What happens if one of the bulbs in a parallel circuit is removed?

_____

Summing Up

1. With what two parts of the bulb does the bulb socket make contact?

_____

_____

2. What do successful arrangements of batteries and bulbs have in common?

_____

_____

3. How do you suppose most of the circuits in your home are wired—in series or in parallel? What is your evidence?

_____

_____

4. How do you suppose automobile headlights are wired—in series or in parallel? What is your evidence?

_____

_____

BATTERIES AND BULBS

CONCEPTUAL PHYSICAL SCIENCE | **Activity**

Magnetism and Electromagnetic Induction | **Patterns of Force**

Seeing Magnetic Fields

Purpose
To explore the patterns of magnetic fields around bar magnets in various configurations

Apparatus
3 bar magnets
iron filings and paper or a magnetic field projectual (iron filings suspended in oil encased in an acrylic envelope)

Discussion
An electric field, as we have learned, surrounds electric charges. In a similar way, a magnetic field surrounds magnets. (Magnetic fields surround other things, too, as we'll learn later.) In this activity, we'll examine the magnetic fields surrounding bar magnets. Although the fields can't be seen directly, their overall shape can be seen by their effect on iron filings.

Procedure

Step 1: Place a bar magnet on a horizontal surface such as your tabletop. Use the iron filings to see the pattern of the magnetic field.

PART A: IRON FILINGS AND PAPER METHOD
Cover the magnet or magnets with a sheet of paper. Then sprinkle iron filings on top of the paper. Jiggle the paper a little bit to help the iron filings find their way into the magnetic field pattern.

PART B: PROJECTUAL METHOD
Step 1: Mix the iron filings by rotating the projectual. Use the glass rod inside the projectual to help stir the iron filings into a fairly even distribution. Hold the projectual upside down for several seconds before placing it on the magnet or magnets. Take care not to scratch the surface of the projectual by moving it across the magnets once it is in place.

Sketch the field for a single bar magnet in Figure 1.

Step 2: Arrange two bar magnets in a line with opposite poles facing each other. Leave about 1 inch between the poles. Use the iron filings to see the pattern of the magnetic field. Sketch the field for opposite poles in Figure 2.

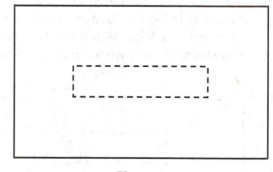

Figure 1

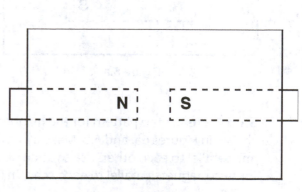

Figure 2

Step 3: Arrange two bar magnets in a line with north poles facing each other. Leave about 1 inch between the poles. Use the iron filings to reveal the magnetic field. Sketch the field for north poles in Figure 3.

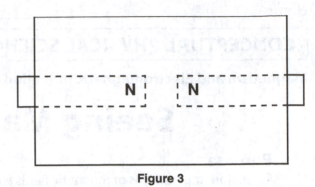

Figure 3

Step 4: Predict the pattern for the magnetic field of two south poles facing each other as shown in Figures 4.a and 4.b. Make a *predictive* sketch in Figure 4.a. Then arrange two bar magnets in a line with south poles facing each other. Leave about 1 inch between the poles. Use the iron filings to reveal the magnetic field. Sketch the *observed* field for south poles in Figure 4.b.

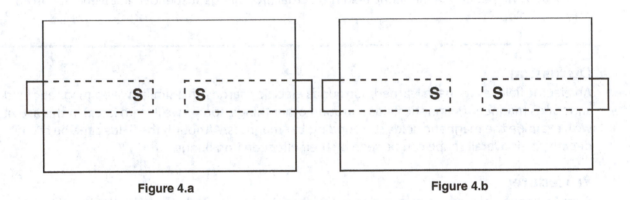

Figure 4.a   **Figure 4.b**

Step 5: Predict the pattern for the magnetic field of two bar magnets parallel to each other as shown in Figures 5.a and 5.b. Make a *predictive* sketch in Figure 5.a. Then arrange two bar magnets parallel to each other. Use the iron filings to reveal the magnetic field. Sketch the *observed* field for two magnets parallel to each other in Figure 5.b.

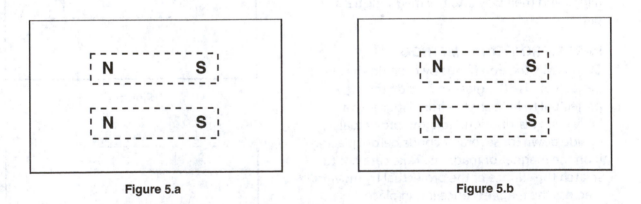

Figure 5.a   **Figure 5.b**

Step 6: Predict the pattern for the magnetic field of two bar magnets antiparallel to each other as shown in Figures 6a. and 6.b. Make a *predictive* sketch in Figure 6.a. Then arrange two bar magnets antiparallel to each other. Use the iron filings to reveal the magnetic field. Sketch the *observed* field for two magnets parallel to each other in Figure 6.b.

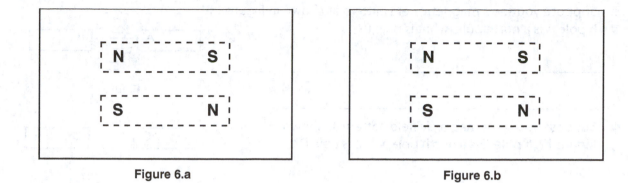

Figure 6.a   Figure 6.b

Step 7: Can you arrange three bar magnets to create the magnetic field shown in Figure 7? If so, how? Is there more than one way to do it?

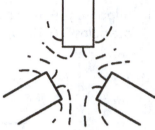

_____

_____

_____

a. Does this pattern show attraction or repulsion?

Figure 7

_____

Step 8: Can you arrange three bar magnets to create the magnetic field shown in Figure 8? If so, how? Is there more than one way to do it?

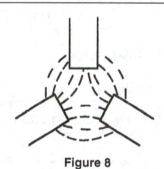

_____

_____

_____

Figure 8

a. Does this pattern show attraction or repulsion?

_____

Summing Up

1. Suppose you see a magnetic field pattern as shown in Figure 9. Can you say for sure which pole is north and which pole is south?

    _____

    _____

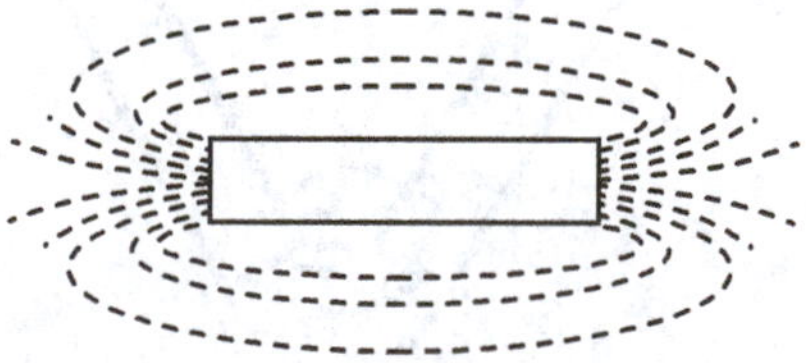

Figure 9

2. Suppose you see a magnetic field pattern as shown in Figure 10. Can you say for sure which pole is north and which pole is south?

    _____

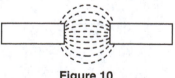

Figure 10

SEEING MAGNETIC FIELDS

3. Suppose you see a magnetic field pattern as shown in Figure 11. If pole A is a north pole, what is pole B?

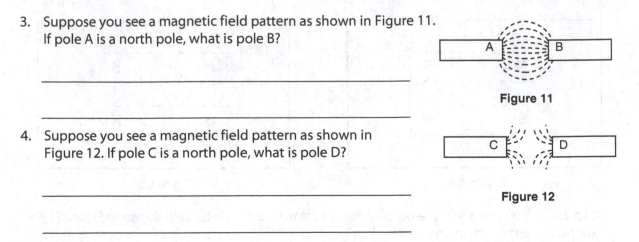

Figure 11

_____

_____

4. Suppose you see a magnetic field pattern as shown in Figure 12. If pole C is a north pole, what is pole D?

Figure 12

_____

_____

5. Suppose you see a magnetic field pattern as shown in Figure 13. If pole E is a north pole, what are poles F, G, H, I, J, K, and L?

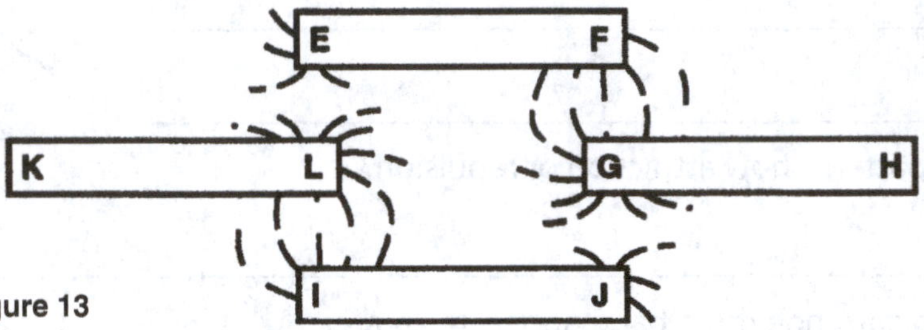

Figure 13

Pole E: North   Pole F:_____   Pole G:_____   Pole H:_____

Pole I:_____   Pole J:_____   Pole K:_____   Pole L:_____

6. Which of the patterns in Figure 14—if either—is/are possible using three bar magnets?

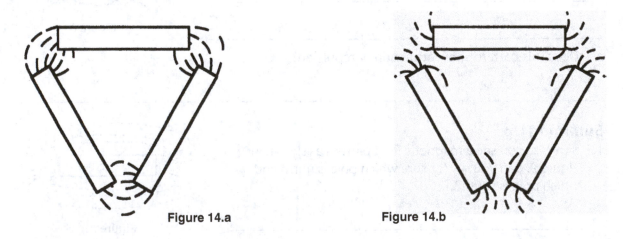

Figure 14.a   Figure 14.b

_____

CONCEPTUAL PHYSICAL SCIENCE	Activity

Magnetism and Electromagnetic Induction   Current and Magnetic Fields

Electric Magnetism

Purpose
To investigate the electric origin of magnetic fields

Apparatus
1 large battery (6-V lantern battery or 1.5-V ignitor battery)
4 small compasses
small platform (a discarded compact disc or equivalent flat object with a hole in the center)
support rod with base
ring clamp
connecting wires

Discussion
In "Seeing Magnetic Fields," you investigated the magnetic fields around various configurations of bar magnets. But where does the magnetic field come from? What's going on inside a bar magnet to make it magnetic? In this activity, you will discover the origin of all magnetic fields.

Procedure
PART A: CURRENT ACROSS A COMPASS
Step 1: Set a compass on your desktop, and allow the needle to settle into its north–south alignment as shown in Figure 1.

Step 2: Stretch a connecting wire across the top of the compass as shown in Figure 2. Rotate the wire clockwise and counterclockwise so that you can see that the wire itself has no effect on the compass needle.

Step 3: Connect the stretched wire to the battery to form a short circuit, and again rotate the wire back and forth as shown in Figure 3. Keep the short circuit connected for no more than 10 seconds.

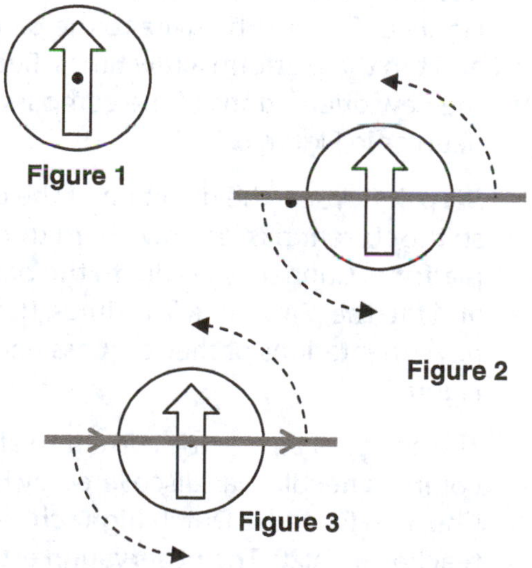

What effect does the current-carrying wire have on the compass?

Figure 1

Figure 2

Figure 3

Step 4: Determine which is more effective in deflecting the compass needle: north–south current or east–west current. Keep in mind that short circuits must not be allowed to run more than 10 seconds and compasses must be level to work properly.

Current has the greatest effect on the compass needle when it runs (select one)

____ north–south.   ____ east–west.

Step 5: Try placing the wire **below** the compass and then running current through it. Does the current affect the needle when the wire passes below the compass?

Step 6. Try reversing the direction of the current by reversing the connections to the battery. What effect does reversing the direction of current have on the deflection of the needle?

PART B: CURRENT THROUGH A PLATFORM OF COMPASSES

Step 1: Arrange the apparatus as shown in Figure 4. Connecting wire passes through the center of the platform. The platform is supported by the ring clamp. The compasses are placed on the platform. Devise a method to have the connecting wire as vertical as possible as it passes through the compass platform.

Step 2: Before running any current through the wire, examine the compass needles by looking down from above the platform. Notice that they all point north, as indicated in Figure 5.

Step 3: Arrange to have current passing upward through the platform as shown in Figure 6. Connect the wire to the battery, and tap the platform a few times. Record the new orientations of the compass needles in Figure 6.

Step 4: Reverse the direction of the current so the current passes downward through the platform. Connect the wire to the battery, and tap the platform a few times. Record the new orientations of the compass needles in Figure 7.

The ability of an electric current to affect a compass needle was discovered by Hans Christian Ørsted, a Dutch high school teacher, in 1820. The observation established the connection between electricity and magnetism. We now know that **all** magnetic fields are the result of moving electric charge (even the magnetic fields of bar magnets). Ørsted's discovery stands as one of the most significant discoveries in the history of physics.

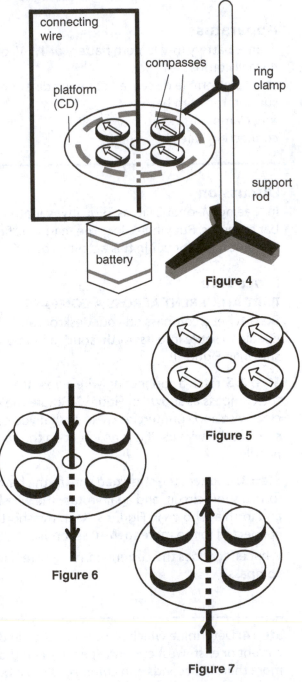

Figure 4

Figure 5

Figure 6

Figure 7

Summing Up

1. Current is passing through the center of a platform that supports four compasses. You are looking straight down at the platform. What is the direction of the current in each configuration shown below: coming toward you or going away from you?

a.   b.

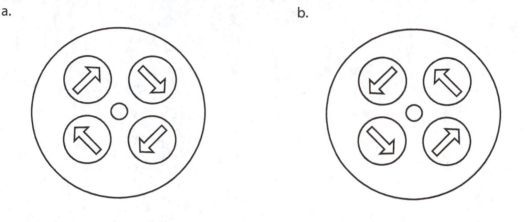

_____   _____

2. The direction of the magnetic field around a wire can be related to the direction of the current in the wire. If you imagine grabbing the wire with your thumb pointing in the direction of the current, your fingers wrap around the wire in the direction of the magnetic field. Which hand must you use for this exercise to give the correct relationship between the direction of the current and the direction of the magnetic field: your left or your right?

_____

3. What is the source of *all* magnetic fields?

_____

CONCEPTUAL PHYSICAL SCIENCE	Activity

Magnetism and Electromagnetic Induction   Simple DC Motors

Motor Madness

Purpose
To investigate the principles that make electric motors possible

Apparatus
handheld generator (Genecon or equivalent)
connecting wires
2 collar hooks (or 2 10-cm lengths of lead-free solder)
about 30 cm of 1/4-inch-diameter wood dowel
support rod with base and rod clamp
2 bar magnets (strong alnico magnets are recommended)

small block of wood (about 2" × 2" × 1")
about 50 cm of lead-free solder
2 rubber bands
2 D-cell batteries

Going Further
St. Louis Motor
6-V lantern battery

Discussion
Perhaps the most important invention of the 19th century was the electric motor. You use a motor whenever you use electric power to make something move. A motor is used to start the engine of a car. Motors are used to spin compact discs. Motors are used to move elevators up and down. A list of motor applications would go on and on. But how do motors turn electric energy into mechanical energy? Let's find out!

An electric current can exert a force on a compass needle (which is simply a small magnet). But Newton's third law of motion suggests that something else must be going on here. What is it? Finish the statement:

If an electric current can exert a force on a magnet, then a magnet

Procedure

PART A: THE MAGNETIC SWING
Step 1: Arrange a solder "swing" by following the instructions below.

a. Make a "sandwich" with the two bar magnets and the wood block, as shown in Figure 1. The magnets must have opposite poles facing each other. Secure the sandwich with the rubber band. See Figure 1.

b. Attach the support rod to the table clamp or ring stand base.

c. Attach the wood dowel to the support rod.

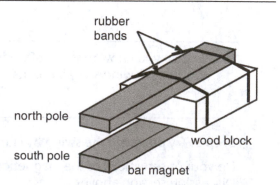

MAGNET: Sandwich a small wood block between two bar magnets. Secure the arrangement with rubber bands as shown. Note that the magnets are antiparallel: opposite poles face each other.

Figure 1

d. Place the collar hooks on the wood dowel about 10 cm apart. (You can use the short lengths of solder to make two hooks if collar hooks are not available.)

e. Bend the long length of solder into a square U-shaped "swing." Make wide hooks at the ends, and hang the swing from the solder hooks on the wood dowel. ***The swing must sway freely on its hooks.*** See Figures 2 and 3.

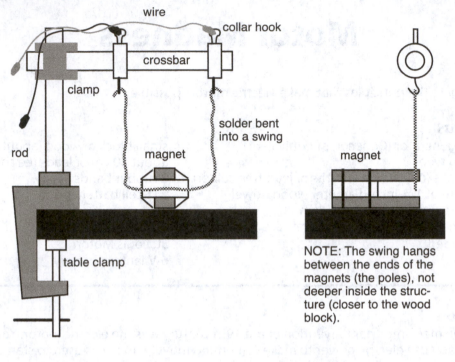

Figure 2. Front View of Arrangement   **Figure 3.** Side View of Arrangement

f. Arrange the height of the crossbar so that the bottom of the solder swing hangs between the magnets.

g. Attach the generator leads to the top leads of the hooks.

Step 2: Crank the generator one way. This will send electric current one way through the swing. Then crank the generator the other way.

a. What effect—if any—does the magnetic field of the bar magnets have on the current in the swing?

_____

b. How do you know that this effect is caused by the current's interaction with the bar magnets? (Would the swing sway if the magnets weren't there?)

_____

Step 3: Try ***pumping*** the swing by cranking the generator back and forth.

Do your observations of the magnetic swing confirm or contradict the prediction stated in the Discussion section above?

_____

PART B: THE SIMPLE MOTOR

Once it was found that a magnetic field could exert a force on an electric current, clever engineers designed practical ways to harness this force. They started to build electric motors. A motor transforms electric energy into mechanical energy. Some simple motors are made of coils of wire and magnets arranged so that when electric current flows through the wires, some part of the motor rotates.

The hand generator you've been using in this and other labs is, in fact, *a motor!*

Step 1: Hold the grip of the generator, but not the crank handle. Touch the two leads of the generator to opposite terminals of a single D-cell battery. What happens?

Step 2: Put two batteries together in series (in a line end-to-end) and touch the leads of the generator to opposite terminals of the arrangement. How is the result different from what happened in Step 1?

Going Further

THE ST. LOUIS MOTOR

Step 1: Obtain a St. Louis Motor and a 6-V lantern battery (or equivalent).

Step 2: Label the following parts of the St. Louis Motor in Figure 4. (Write the names of the parts, not the single-letter designations from the list!)

a. **Armature:** coiled copper wire that will carry current when the battery is connected. Current in the armature produces a magnetic field that interacts with the field magnets.

b. **Field magnets:** permanent magnets whose fields will interact with the magnetic field in the armature.

c. **Split-ring commutator:** metal cylinder on the spinning axle of the motor. Notice the gaps in the metal (the "split" in the ring). The split-ring commutator directs the current through the armature so that the armature continues to spin rather than getting stuck in one position.

d. **Commutator bridge:** the yolk that connects the brushes. The angle of the commutator bridge is adjustable.

e. **Brushes:** metal structures that touch the split-ring commutator. They are connected to the commutator bridge.

f. **Terminals:** electrical connection posts on the commutator bridge.

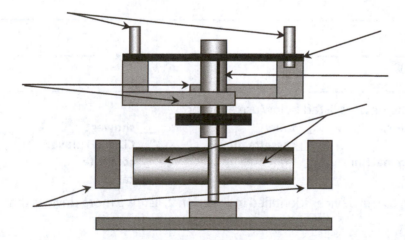

Figure 4. St. Louis Motor

SAFETY CAUTION: When the St. Louis Motor is running, the armature can spin at very high speed. Handle the motor with caution, and keep it away from loose clothing, dangling jewelry, long hair, and so on. Do not attempt to stop the spinning armature with fingers, pencils, or the like, because injury will result.

Step 3: Connect the 6-V lantern battery to the St. Louis Motor. If the motor does not immediately spin upon connection, troubleshoot as follows:

First: Jiggle or tap the St. Louis Motor housing or base.

Second: Give the armature a spin in one direction. If that doesn't work, try spinning the armature the other way.

Third: Disconnect the battery, and adjust the brushes so that they are in good contact with the split-ring commutator. Reconnect the battery and try again.

Fourth: Ask your instructor for assistance.

Step 4: Adjust the angle of the commutator bridge, and record your findings.

_____

_____

Step 5: Find the optimal angle for the commutator bridge. Adjust the position of the field magnets, and record your findings. If possible, remove the field magnets completely. Then try reversing the polarity of one so that north faces north or south faces south.

_____

_____

Step 6: Your battery provides a relatively constant amount of electrical power. What do you think would happen if you could increase the electrical power supplied to the motor?

_____

Summing Up

1. The magnetic swing swayed as a consequence of interaction between the current in the wire and the magnetic field of the bar magnets. What are some ways in which this force could be made stronger (and thereby push the swing farther in or farther out)?

_____

_____

_____

2. Which of the devices listed below use a motor?

___alarm clock	___toilet	___shower	___blow dryer
___shaver	___cassette player	___CD/DVD player	___radio
___vending machine	___light bulb	___computer	___TV
___VCR	___washing machine	___car	

3. In addition to the devices identified in question 2, list two more devices that use motors.

_____

_____

CONCEPTUAL PHYSICAL SCIENCE	**Activity**

Generator Activator

Purpose
To investigate electromagnetic induction, the principle behind electric generators

Apparatus
bar magnet (strong alnico magnet is recommended)
air core solenoid or a long wire coiled into many loops
connecting wires
galvanometer (–500 µA – 0 – +500 µA recommended)
handheld generator (Genecon or equivalent)
access to a second handheld generator

Discussion
In 1820, Hans Christian Ørsted found that electricity could create magnetism. Scientists were convinced that if electricity could create magnetism, then magnetism could create electricity. Still, 11 years would pass before the induction of electricity from magnetism would be discovered and understood. The best minds of the day set out to make this widely anticipated discovery, but it was Michael Faraday who put all the pieces together. In this activity, you will use a magnet to create an electric current and see how this effect is applied in electric generators.

Procedure

PART A: ELECTROMAGNETIC INDUCTION
Step 1: Connect the galvanometer to the coil (air core solenoid or looped wire).

Step 2: Determine a method for producing current in the coil using the bar magnet.

a. Describe your findings.

_____

_____

b. Can current be produced if
   i. the coil is at rest? If so, how?

_____

   ii. the magnet is at rest? If so, how?

_____

   iii. both the coil and the magnet are at rest? If so, how?

_____

PART B: THE GENERATOR

Once it was found that a changing magnetic field could induce an electric current, clever engineers figured out practical ways to harness induction. They started to build electric generators. A generator transforms mechanical energy into electric energy. Some simple generators are made of coils of wire and magnets arranged so that when some part of the generator is rotated, electric current moves through the wires.

The hand-crank generator you have used in previous labs is such a device.

Step 1: Disconnect the coil from the galvanometer, and attach the leads of the hand-crank generator to the galvanometer. Slowly turn the handle until the meter responds.

a. What does the galvanometer show?

_____

b. What happens if you turn the handle the other way?

_____

The invention of the generator made it possible to produce continuously flowing electric energy without the use of chemical batteries. Further developments led to the wide-scale distribution of electric energy and the availability of household electricity.

Large-scale electrical distribution grids are powered by large-scale generators. These generators are commonly powered by steam turbines. The heat used to generate the steam is typically produced by burning coal or oil, or as a by-product of controlled nuclear reactions.

Step 2: Use a generator to power a motor! Connect two hand-crank generators to each other. Crank the handle of one of the generators.

a. What happens to the handle of the other generator? Would you say the other generator is acting as a motor?

_____

b. Not all the energy you put into the generator turns into mechanical energy in the motor. What is your evidence of this?

_____

Summing Up

1. Name each device described below.

   a. **Transforms electric energy into mechanical energy:** _____

   b. **Transforms chemical energy into electric energy:** _____

   c. **Transforms mechanical energy into electric energy:** _____

2. What happens to the energy lost between the generator and the motor in Step 2 of Part B above?

_____

3. A classmate suggests that a generator could be used to power a motor that could then be used to power the generator. What do you think about this proposal, and why?

_____

_____

CONCEPTUAL PHYSICAL SCIENCE	Experiment
Magnetism and Electromagnetic Induction	Electromagnetism Simulation

Faraday's Electromagnetic Lab

Purpose
To manipulate simulated magnets, compasses, and coils to see how magnetic fields interact with electric currents

Apparatus
computer
PhET sim "Faraday's Electromagnetic Lab" (available at http://phet.colorado.edu)

Discussion
When Hans Christian Ørsted discovered that electricity could be used to produce magnetism, the scientific community anticipated that it wouldn't be long before someone would discover how magnetism could be used to produce electricity. But more than ten years would pass before Michael Faraday solved the puzzle.

The application of engineering to electromagnetism led to motors and generators. Nearly every electrical device that produces motion uses a motor. Any device that is plugged into a wall outlet draws power from a generator. Our reliance on applications of electromagnetism is never more apparent than during a power outage.

The interactions between electricity and magnetism are not always easy to grasp. In this activity, you will manipulate elements in a simulated laboratory and get visual feedback.

Procedure
PART A: BAR MAGNET
Step 1: Run the PhET sim "Faraday's Electromagnetic Lab." It should open to the Bar Magnet tab. Maximize the window. You should see a bar magnet, a compass, and a compass needle grid.

Step 2: Center the bar magnet horizontally on the fourth or fifth row from the top. Set the large compass just below the bar magnet at its midpoint. It's okay for the two objects to be touching. See Figure 1.

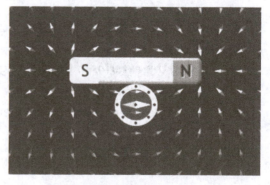

Figure 1

Step 3: If the compass needles (in the grid or in the large compass) are to be thought of as arrows indicating the direction of the bar magnet's magnetic field, each one should be visualized as pointing

___"redward"   ___"whiteward".

Step 4: Using the on-screen slider in the control panel, run the strength of the bar magnet up and down. How does the sim show the difference between a strong magnet and a weak magnet?

_____

_____

Step 5: How does the strength of the magnetic field change with increasing distance from the bar magnet, and how does the sim show this?

_____

_____

Step 6: With the magnet at its strongest, reverse its polarity using the on-screen "Flip Polarity" button in the control panel. What are the ways in which the sim reflects this polarity reversal?

_____

_____

Step 7: Describe the behavior of the compass during a polarity reversal (magnet initially at 100%)
a.   when the compass is touching the bar magnet at its midpoint.

_____

_____

b.   when the compass is far from the bar magnet (touching the bottom of the sim window), but still on a perpendicular bisector of the bar magnet.

_____

_____

c.   when the compass is far from the bar magnet and the magnet's strength is set to 10%.

_____

_____

Step 8:
a.   Around the **_exterior_** of the bar magnet, the direction of the magnetic field is from its _____ pole to its _____ pole.

b.   What is the direction of the magnetic field in the **_interior_** of the bar magnet? And how did you find out?

_____

_____

PART B: ELECTROMAGNET

Step 1: Run the PhET sim "Faraday's Electromagnetic Lab." Maximize the window. Click the on-screen Electromagnet tab. Arrange the on-screen elements so that the top of the battery is along the second or third row of the compass grid. Notice that the magnetic field around the coil is very similar to the magnetic field around the bar magnet.

Step 2: There is no "Strength %" slider on the control panel.
a.   How can you change the strength of the electromagnet?

_____

_____

b.   In real life, is it easier to change the strength of a bar magnet or of an electromagnet?

_____

_____

Step 3: There is no "Flip Polarity" button on the control panel. How can you reverse the polarity of the electromagnet?

_____

_____

Step 4: In the control panel, switch the Current Source from the battery (DC: direct current) to an oscillator (AC: alternating current). If necessary, move the electromagnet so that you can see the entire oscillator.

a.   What does the vertical slider on the AC source do?

_____

_____

b.   What does the horizontal slider on the AC source do?

_____

_____

Step 5: What should the sliders be set to in order to create a "dance party" display? Can you make the dance party even more annoying using the Options menu? Describe.

_____

_____

PART C: PICKUP COIL

Step 1: Run the PhET sim "Faraday's Electromagnetic Lab." Maximize the window. Click the Pickup Coil tab. You should see a bar magnet, a compass needle grid, and a coil attached to a light bulb.

Step 2: Describe the most effective way of using the magnet and the coil to light the bulb if

a. the coil cannot be moved.

_____

b. the magnet cannot be moved.

_____

Step 3: Rank the arrangements and motions shown below from most effective to least effective in terms of lighting the bulb, allowing for ties. For example, if A were most effective, B were least effective, and C and D were equivalent to one another, the ranking would be A > C = D > B.

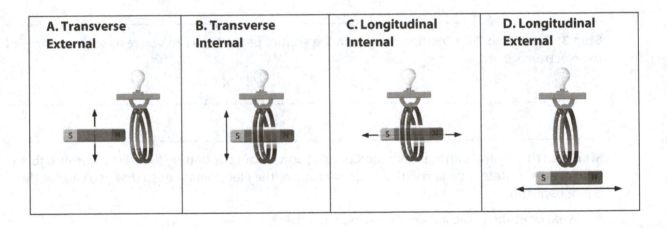

| A. Transverse External | B. Transverse Internal | C. Longitudinal Internal | D. Longitudinal External |

_____

Step 4: Move the bar magnet through the coil, and observe the motion of the electrons in the forward arc of the coil loops. Report the correlations between magnet motion and electron motion.

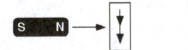 

a. Magnet approaches from the left, north pole first; electrons move downward.

b. Magnet departs to the right, south end last; electrons move upward.

c. Magnet approaches from the right, south pole first; electrons move _?_.

d. Magnet departs to the left, north end last; electrons move _?_.

e. Magnet approaches from the left, south pole first; electrons move _?_.

f. Magnet departs to the right, north end last; electrons move _?_.

g. Magnet approaches from the right, north pole first; electrons move _?_.

h. Magnet departs to the left, south end last; electrons move _?_.

PART D: TRANSFORMER

1. Run the PhET sim "Faraday's Electromagnetic Lab." Maximize the window. Click the on-screen Transformer tab. You should see an electromagnet and a pickup coil.

2. Experiment with the various control panel settings, and with the positions of the electromagnet and the pickup coil, to determine a method for getting the most light out of the bulb. Describe the settings and locations.

_____

_____

_____

_____

_____

PART E: GENERATOR

1. Run the PhET sim "Faraday's Electromagnetic Lab." Maximize the window. Click the on-screen Generator tab. You should see a faucet, a paddlewheel with a bar magnet, a compass, and a pickup coil.

2. Experiment with the various settings to determine a method for getting the most light out of the bulb. Describe the settings.

_____

_____

_____

_____

_____

3. What is the story of light production here? Organize and connect the given "plot elements," and add any key elements that were omitted from the list to construct the complete story.

- *light radiated from the bulb*
- *motion of the bar magnet*
- *changing magnetic field*
- *kinetic energy of the water*
- *induced electric current*
- *heat the filament*

_____

_____

_____

_____

_____

_____

_____

_____

CONCEPTUAL PHYSICAL SCIENCE	Activity

Waves and Sound

Slow-Motion Wobbler

Purpose

To observe and explore the oscillation of a tuning fork

Apparatus

low-frequency tuning fork (40–150-Hz forks work best for large amplitudes)
tuning fork mallet (or equivalent, optional)
bright strobe light with a widely variable frequency (inexpensive, party strobe lights **won't** do)
beaker (any size)
access to water

Discussion

The tines of a tuning fork oscillate at a very precise frequency. That's why musicians use them to tune instruments. In this activity, you will investigate their motion with a special illumination system—a **stroboscope.**

Procedure

Step 1: Strike a tuning fork with a mallet or the heel of your shoe (do **not** strike it against the table or other hard object). Does it appear to vibrate? Try it again, this time dipping the tip of the tines just below the surface of water in a beaker. What do you observe?

_____

_____

Step 2: Now dim the room lights, and strike a tuning fork while it is illuminated with a strobe light. For best effect, use the tuning fork with the longest tines available. Adjust the frequency of the strobe so that the tines of the tuning fork appear to be stationary. Then carefully adjust the strobe so that the tines slowly move back and forth.

a.   Describe your observations.

_____

_____

_____

_____

b. Which illustration below best depicts the motion of the tuning fork tines? Circle the correct sequence; cross out the incorrect sequence.

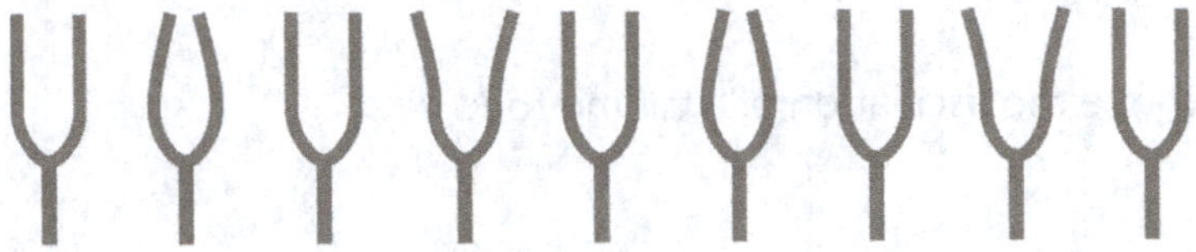

Step 3: Strike a tuning fork and observe how long it vibrates. Repeat, placing the handle against the tabletop or counter.

a. Is the sound louder or quieter?

_____

b. Does the *time* the fork vibrates increase or decrease? How does this make sense in terms of energy?

_____

_____

_____

Summing Up

1. What happens to the air next to the tines as they oscillate?

_____

_____

2. What would happen if you struck the tuning fork in outer space?

_____

_____

_____

CONCEPTUAL PHYSICAL SCIENCE	Activity

Waves and Sound   Wave Mechanics Simulation

Water Waves in an Electric Sink

Purpose
To observe and control waves in a ripple tank simulation to learn the basics of wave mechanics

Apparatus
computer
PhET simulation "Wave Interference" (available at http://phet.colorado.edu)

Discussion
The ripple tank is an effective (though cumbersome) classroom device used for demonstrating and exploring wave phenomena. A simple version is shown on page 254 in your text. More elaborate ones resemble a small glass table with raised edges. Water is poured onto the table and kept from spilling by the raised edges. Typically, a strong point light source is placed above the tank, and shadows of ripples can be seen below the tank. A small ball attached to a motor bobs in and out of the water to make waves with consistent amplitude and wavelength.

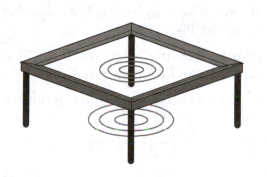

Figure 1. A Ripple Tank

A variety of wave phenomena can be demonstrated using the ripple tank. This activity uses a ripple tank simulation, so you'll be able to investigate waves without the water.

Procedure

PART A: CRESTS AND TROUGHS

Step 1: When the simulation opens, you will see a faucet dripping water into a large sink. The drops create ripples in the water in the sink.

Step 2: Locate the "Rotate View" slider in the control panel on the right side of the window. Drag the slider to the right. Doing so rotates your view of the sink from a top view to a side view.

Step 3: Locate the "Pause" button at the bottom of the window. Try to pause the animation when the water under the faucet rises to its highest point (close to or touching the faucet itself).

Step 4: Locate the "Show Graph" button below the blue water of the ripple tank. Click it to activate the graph. Notice that the graph and the side view of the water match each other.

Step 5: Slide the "Rotate View" slider back to the left so that it shows the top view of the water.

a.   In the spaces below, sketch the wave pattern as seen from the top and from the side.

Side View (Graph)	Top View

b.   In both views (side view and top view), label a **crest** and a **trough**.

c.   In both views, label one wavelength.

PART B: AMPLITUDE

Step 1: Pause the animation. Locate the frequency slider below the faucet. Set the frequency to its maximum value by moving the slider all the way to the right. Restart the animation by clicking the on-screen "Play" button.

Step 2: Locate the "Amplitude" slider. Slide it to various positions (to the left and right), and observe the effect this has on the simulation.

a.   Does a change in amplitude result in a change in the size of the water drops? If so, how?

_____

_____

b.   How are high-amplitude waves different from low-amplitude waves?

_____

_____

c.   Review your sketches above (side view and top view) of the wave. Label the amplitude of the wave.

d.   Which view—side or top—is better suited for labeling the amplitude? Explain?

_____

_____

e.   What—if anything—happens to the amplitude of each wave as it gets farther away from the source?

_____

_____

PART C: FREQUENCY

Step 1: Pause the animation. Set the amplitude to its maximum value by moving the slider all the way to the right. Restart the animation by clicking the on-screen "Play" button.

Step 2: Move the "Frequency" slider to various positions (to the left and right), and observe the effect this has on the simulation.

a.   How are high-frequency waves different from low-frequency waves? (What *is* different?)

_____

_____

b.   How are high-frequency waves the same as low-frequency waves? (What *isn't* different?)

_____

_____

c.   Two students disagree about an observed difference between high-frequency waves and low-frequency waves. One says high-frequency waves are faster than low-frequency waves; the other claims both waves have the same speed. What do you think?

_____

_____

_____

_____

d.   What is the relationship between the frequency (*f*) of the wave source (the dripping faucet) and the wavelength (λ of the waves?
___ *Direct proportionality: λ ~ f. The wavelength increases as the frequency increases.*
___ *Inverse proportionality: λ ~ 1/f. The wavelength increases as the frequency decreases.*
___ *No apparent relationship. The wavelength doesn't appear to be related to the frequency.*

e.   What—if anything—happens to the frequency of each wave as it gets farther away from the source?

_____

_____

Summing Up

1.   Examine the illustrations below. Each represents a ripple tank wave. Some are side views; some are top views. Describe the amplitude of the wave and the frequency of its source by using the terms "high" and "low." Please examine all the patterns before recording your descriptions. (*Hint:* Waves a–d are all different from one another.)

a.   _____ **amplitude**

_____ **frequency**

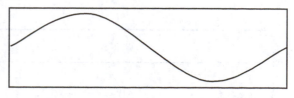

b. _____ amplitude
_____ frequency

c. _____ amplitude
_____ frequency

d. _____ amplitude
_____ frequency

2. What single aspect of a wave does its amplitude best represent?

___ **speed** ___ **wavelength** ___ **frequency** ___ **energy** ___ **period**

3. a. Which control on a music player or television set enables you to increase or decrease the amplitude of the sound waves that come out of it?

_____

Recall what happens to the amplitude of a wave as the wave gets farther from the source. Imagine a portable music player playing music in a large, open field. At some distance from the player, the amplitude of the sound waves diminishes to zero, and the sound cannot be heard. Consider the three-dimensional space in which the sound can be heard.

b. How might you increase that space, and what is three-dimensional space called (in geometry)?

_____

_____

c. Ripple tanks are used to observe two-dimensional waves. What should be the name of the amplitude control for two-dimensional waves?

_____

_____

CONCEPTUAL PHYSICAL SCIENCE	Activity

Waves and Sound   **Sound Wave Manipulation and Interpretation**

High Quiet Low Loud

Purpose
To use sound generation software to manipulate the amplitude and wavelength of a sound wave, and observe the connections between the plot of a sound wave and the characteristics of the sound

Apparatus
computer
sound generation software (Pasco's DataStudio with Waveport or equivalent)
headphones (optional: in-line volume control)
signal splitters (if two or more lab partners are connected to a single computer)

Discussion
Sound is a mechanical wave. It is a longitudinal wave that travels through materials. Sound is generated by a vibrating source and is received by objects that can also vibrate, such as the tympanic membrane (the eardrum). Like all mechanical waves, sound waves have amplitude and wavelength. Your sense of hearing is able to distinguish sound waves by their tone (or pitch) and volume (or loudness). Wavelength and amplitude are related to volume and tone. In this activity, you will find the specific connections.

Procedure
Step 1: If it's not already running, start the computer.

Step 2: While the computer is starting, connect the headphones to the signal splitters as needed. Every member of the lab group should have his or her own pair of headphones.

Step 3: Make sure the computer's sound is on and turned all the way up.

Step 4: Start the sound generation software (for example, DataStudio with Waveport).

Step 5: Activate the sound-generating component of the software (for example, SoundCreation). If the software has a separate analysis component (such as SoundAnalyzer), close it for now. If you have difficulty, ask your instructor for assistance.

Step 6: Start the sound by, for example, clicking the on-screen image of the speaker. Examine the initial sound as graphically displayed by the software.

Notice that the **_longitudinal_** sound wave is represented here as a **_transverse_** wave. The displayed wave actually represents the electrical signal used to drive the speaker. Current in one direction pushes the speaker cone out; current in the other direction pulls the speaker cone in. The sinusoidal current pushes and pulls the speaker cone many times each second to create sound waves in the air.

Any changes made to the electrical signal will change the resulting sound.

Step 7: Make the wave *smaller or taller* by altering the amplitude of the wave. There should be an on-screen control (such as a hand icon) that you can move by clicking and dragging.

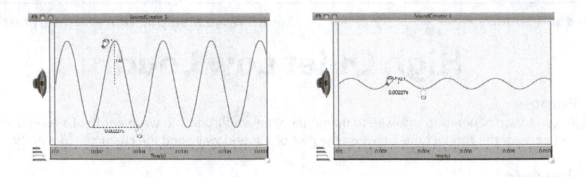

Figure 1. Making the wave taller or smaller by manipulating the amplitude.

What effect does changing the signal's amplitude have on the sound that you hear? Be specific: What does increasing the amplitude do, and what does decreasing the amplitude do?

_____

_____

_____

Step 8: Make the wave *longer or shorter* by altering the wavelength of the wave. There should be an on-screen control (such as a hand icon) that you can move by clicking and dragging.

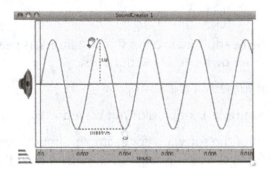 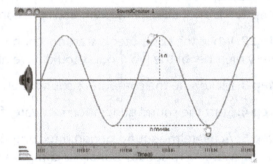

Figure 2. Making the wave longer or shorter by manipulating the wavelength.

What effect does changing the signal's wavelength have on the sound that you hear? Be specific: What does increasing the wavelength do, and what does decreasing the wavelength do?

_____

_____

_____

Summing Up

1. Consider the first wave to be a wave with a medium volume (loudness) and average tone (pitch).

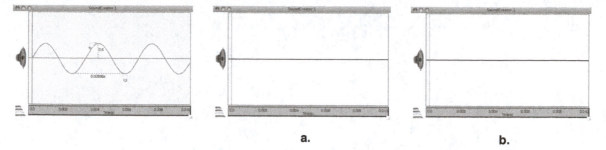

a.

b.

Draw:
a. the signal for a louder-sounding wave with roughly the same pitch.
b. the signal for a quieter-sounding (but not silent) wave with roughly the same pitch.

2. Consider the first wave to be a wave with a medium volume (loudness) and average tone (pitch).

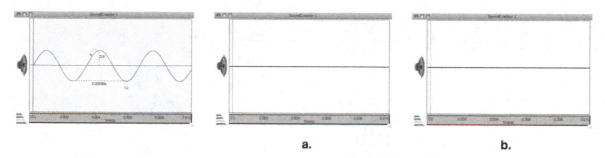

a.

b.

Draw:
a. the signal for a higher-pitched wave with roughly the same volume.
b. the signal for a lower-pitched wave with roughly the same volume.

3. The waves illustrated below represent signals for speakers.

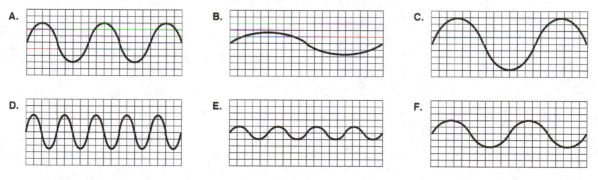

a. Rank them from loudest to quietest.

_____

b. Rank them from highest to lowest.

_____

HIGH QUIET LOW LOUD

CONCEPTUAL PHYSICAL SCIENCE	Activity

Waves and Sound   **The Products of Sound Wave Interference**

Wah-Wahs and Touch-Tones

Purpose
To use sound generation software to produce and observe the interference of multiple sound waves

Apparatus
computer
sound generation software (Pasco's DataStudio with Waveport or equivalent)
headphones (optional: in-line volume control)
signal splitters (if two or more lab partners are connected to a single computer)

Going Further
access to the Internet

Discussion
When two sound waves interfere with one another, the resulting sound pattern is referred to as **beats.** The beat pattern is determined by the waves that produce it. Piano tuners use their knowledge of beats to tune pianos. A tuning fork with a key's frequency is struck, and the note is played. If beats are heard, an adjustment needs to be made. Land-line telephones send tones when keys are pressed. The tones are combinations of two notes played at the same time. Beats are all around us!

Procedure
SETUP
Step 1: If it's not already running, start the computer.

Step 2: While the computer is starting, connect the headphones to the signal splitters as needed. Every member of the lab group should have his or her own pair of headphones.

Step 3: Make sure the computer's sound is on and turned all the way up.

Step 4: Start the sound generation software (for example, DataStudio with Waveport).

Step 5: Activate the sound-generating component of the software (for example, SoundCreator). If the software has a separate analysis component (such as SoundAnalyzer), close it for now. If you have difficulty, ask your instructor for assistance.

Step 6: Start the sound by, for example, clicking the on-screen image of the speaker. Consider the image below to represent the initial sound as displayed by the software. Any changes made to the electrical signal will change the resulting sound.

Step 7: Double-click in the SoundCreator window to activate SoundCreator's Settings window.
a.   The on-screen Appearance Pane should open by default. Type a "2" into the Simultaneous Tones box (1 is the default value).
b.   Click the Tones Display tab. Click the checkbox to activate the Phase Tool. Click the checkbox to Show Frequency Buttons. Click OK to close the Settings window.

PART A: PHASE

Step 1: Click the on-screen speaker icon in the lower window to activate the tone.

The tone consists of two input waves. The two input waves are depicted in the upper window. One is red; one is green. Colored buttons to the left of the waveforms enable the user to select an input wave.

The result of the two waves is depicted in the lower window. The two colored dots near the speaker icon show the input waves. Arrows from the dots to the speaker indicate which input waves are active.

Step 2: In the lower window, click the green dot to deactivate the green input wave.

What effect—if any—does removing one input wave have on the resulting sound?

_____

_____

Step 3: Use the phase tool (the on-screen hand icon farthest to the left in the upper window) to change the phase of the red input wave. Adjust the phase forward and back, while listening to the tone.

What effect—if any—does changing the phase of a wave have on the resulting sound?

_____

_____

PART B: INTERFERENCE

Step 1: Return the phase of the red wave back to its original state. Reactivate the green wave by clicking the green dot near the speaker. The lower window and the resulting sound should reflect that there are now two input waves.

Step 2: Now adjust the phase of the green wave in the upper window, and observe the resulting sound.

Describe, using words and pictures, the result when the two waves are completely out of phase. Figure 1.a shows the result when the waves are completely in phase. Figure 1.b shows the two input waves completely out of phase (but it does not show the result).

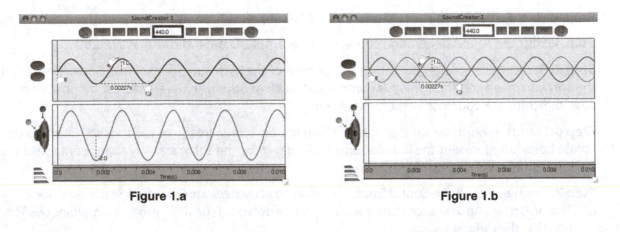

Figure 1.a   Figure 1.b

_____

_____

PART C: BEATS

Step 1: Return the waves to their original state (in phase).

If you ever have trouble resetting the input waves to their original state, close the SoundCreator window and create a new one. Then adjust the settings as described in Steps 1–7.

Step 2: Drag the on-screen wavelength tool (hand icon connected to the trough) to adjust the frequency of one of the input waves.

What is the effect when two input waves have slightly different frequencies?

_____

_____

Step 3: Using the frequency control buttons, adjust one input wave to have a frequency of 440.0 Hz and the other to have a frequency of 441.0 Hz.

Step 4: Listen for the beat frequency. The beat frequency is equal to the difference between the frequencies of the input waves. In this case, the beat frequency is 441.0 Hz – 440.0 Hz = 1.0 Hz. You should hear a 1.0-Hz "wah-wah" pattern in the resulting sound.

How else can you generate a 1.0-Hz beat frequency without changing the frequency of the 440.0-Hz input wave?

_____

_____

Step 5: Experiment with other combinations of input wave frequencies to create beats. Low frequencies produce interesting results.

Summing Up

1. How are the terms *constructive interference* and *destructive interference* related to the relative phase of the two input waves?

   When two waves are in phase, _____ *interference occurs.*

   When two waves are out of phase, _____ *interference occurs.*

2. What is the beat frequency when the adjacent piano notes C (262 Hz) and C-sharp (277 Hz) are played simultaneously? Show your work.

   _____

   _____

Going Further: Touch-Tones

Dual-tone multifrequency (DTMF) phone tones use two frequencies for each key. For example, when the "1" key is pressed, the tone that plays is the combination of 697 Hz and 1209 Hz. The resulting beat frequency is 1209 Hz – 697 Hz = 512 Hz.

1. Access the Internet to research the frequency pairs for all the other keys on a telephone keypad.

   1 = 697 + 1209   2 = _____ + _____   3 = _____ + _____

   4 = _____ + _____   5 = _____ + _____   6 = _____ + _____

   7 = _____ + _____   8 = _____ + _____   9 = _____ + _____

   * = _____ + _____   0 = _____ + _____   # = _____ + _____

2. Determine the beat frequency for each key.

   1 = 512 Hz   2 = _____   3 = _____

   4 = _____   5 = _____   6 = _____

   7 = _____   8 = _____   9 = _____

   * = _____   0 = _____   # = _____

CONCEPTUAL PHYSICAL SCIENCE	Activity

Light   **Plane Mirror Images**

Mirror rorriM

Purpose
To investigate the minimum size of a mirror in which you want to see a full image of yourself

Apparatus
large mirror, preferably full-length
ruler
masking tape

Discussion
Why do shoe stores and clothing shops have full-length mirrors? Must a mirror be as tall and as wide as you for you to see a complete image of yourself?

Procedure
Step 1: Stand about an arm's length in front of a vertical full-length mirror. Reach out and place a small piece of masking tape on the image of the top of your head. Now stare at your toes. Place the other piece of tape on the mirror where your toes are seen. Use a meter stick to measure the mirror's distance from the top of your head to your toes, and then measure your actual height. How does the distance between the pieces of tape on the mirror compare with your height?

Step 2: Now stand about 3 meters from the mirror and repeat Step 1. Stare at the top of your head and toes, and have an assistant move the tape so that the pieces of tape mark where your head and feet are seen. Move farther away or closer, and repeat. What do you discover?

Summing Up
1. Does the location of the tape depend on your distance from the mirror?

_____

_____

2. What is the shortest mirror you can use to see your entire image? Do you **believe** it?

_____

_____

Going Further

Try this one if a full-length mirror is not readily available **or** you are a disbeliever! Hold a ruler next to your eye. Measure the height of a common pocket mirror. Hold the mirror in front of you so that the image includes the ruler. How many centimeters of the ruler appear in the image? How does this compare with the height of the mirror?

_____

_____

_____

Name _____ Section _____ Date _____

CONCEPTUAL PHYSICAL SCIENCE	**Activity**

Light

Color Mixing in Digital Displays

Pixel Peeping

Purpose
To use a magnifier to examine the pixels of a computer display and see how the display's colors are created

Apparatus
__computer with display (monitor or screen)
__PhET sim "Color Vision" (available at http://phet.colorado.edu)
__overhead transparency sheet (or equivalent)
__small magnifier (30× or 60×–100×, for example)

Discussion
The term "pixel peeper" is used among digital photographers to describe someone more interested in the technical specifications of camera gear than in photography itself. But in this activity, we will literally examine ("peep") at the individual picture elements ("pixels") of a computer display to see how colors are created. The results might surprise you!

Procedure
Step 1: Place the transparency over the screen to protect it during this activity. With laptops, the transparency should rest in place without any further need to secure it. Ask your instructor for assistance if you have difficulty with the transparency sheet.

Step 2: Launch the PhET sim "Color Vision."

Step 3: While in the RGB Bulbs tab, use the on-screen slide control to increase the red light to maximum.

Step 4: Place the small magnifier onto the on-screen character's "perception bubble" to see what red looks like up close. (Recommendation: 60× is plenty of magnification; place the magnifier on the transparency on the section of screen to be examined, and focus it there.) Describe what you see using words and pictures.

_____

_____

_____

Step 5: Slide the red control back to zero, and increase the green to maximum. Use the magnifier to see what green looks like up close. Describe what you see using words and pictures.

_____

_____

_____

Step 6: Slide the green control back to zero, and increase the blue to maximum. Use the magnifier to see what blue looks like up close. Describe what you see using words and pictures.

_____

_____

_____

Step 7: Switch to the Single Bulb tab. Switch the filter to "Off." Use the magnifier to see what white looks like up close. Describe what you see using words and pictures.

_____

_____

_____

Step 8: Switch the color filter to "On." Set the Filter Color to as pure a yellow as you can (this may be the default color that the filter is set to when switched on).
a. Prediction. What will yellow look like up close?

_____

_____

_____

b. Observation. What does yellow look like up close? Describe what you see using words and pictures.

_____

_____

_____

Step 9: Try watching through the magnifier while the Filter Color is adjusted from one extreme to another. What do you see through the magnifier at each of the spectral colors?

Red

_____

_____

Orange

_____

_____

Yellow

_____

_____

Green

_____

_____

Blue

_____

_____

Violet

_____

_____

Step 10: Return the sim to the RGB Bulbs tab. Knowing what you know about yellow, how can you make the "perception bubble" bright yellow?

_____

_____

_____

Summing Up

1. The term "pixel" is short for "picture element." A computer screen is an array of pixels. Standard Definition is 640 × 480 pixels, High Definition is 1280 × 720 pixels, Full HD is 1920 × 1080 pixels, 4K UHD is 3840 × 2160 pixels, and 8K UHD is 7680 × 4320 pixels. Each pixel is made of phosphors. A single phosphor emits one color of light. How many phosphors (tiny color lights) are there in each pixel?

_____

2. Which colors correspond to the individual phosphors within a pixel? *Note*: Some of the color names listed below may be unfamiliar to you. If the computer you're using is connected to the Internet, you might search the name of the color and then refine your search to produce images (rather than web links).

__blue (pure)	__brown	__chartreuse	__cyan	__gray
__green (pure)	__magenta	__mint	__orange	__periwinkle
__red (pure)	__violet	__white	__yellow	__cerulean blue

3. Arrange the following and describe the resulting color (in one word, if possible; use the list above as a guide). *Note*: 1.0R means red at 100% intensity; 0.5G means green at 50% intensity.

a. 1.0R + 1.0G + 1.0B = _____ 　　b. 1.0R + 1.0G = _____

c. 1.0R + 1.0B = _____ 　　d. 1.0G + 1.0B = _____

e. 1.0R + 0.5G = _____ 　　f. 0.5R + 1.0G = _____

g. 1.0R + 0.5B = _____ 　　h. 0.5R + 1.0B = _____

i. 1.0G + 0.5B = _____ 　　j. 0.5G + 1.0B = _____

k. 0.5R + 0.5G + 0.5B = _____

Going Further

Examine the pixel structure of other items (such as smart phones, plasma TVs, tablet computers, and the like). Are they all the same, or do they differ? What—if anything—do they have in common? Describe using words and pictures.

Device: _____

Pixel structure: _____

_____

_____

Device: _____

Pixel structure: _____

_____

_____

Device: _____

Pixel structure: _____

_____

_____

CONCEPTUAL PHYSICAL SCIENCE	Activity

Light   **Resonance and Scattering**

Why the Sky Is Blue

Purpose
To see why the daytime sky is blue while the sunrise/sunset sky is red

Apparatus
a pair of resonant (sympathetic) tuning forks: matching tuning forks attached to wood sound boxes
tuning fork mallet
Laser Viewing Tank (or equivalent)
small, bright flashlight (such as LED Mini Maglite)
access to water
scattering agent
stirring rod
paper towel

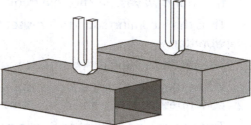

Discussion
In Chapter 11 of your text, you read about why the sky is blue. In this activity, we extend the analogy of sound scattered by tuning forks to light scattered by **optical tuning forks**, the particles that make up the air.

BLUE SKY INGREDIENTS
Earth's sky is made of air. Air is mostly molecular nitrogen (N_2) and molecular oxygen (O_2). Nitrogen and oxygen are neighbors on the periodic table; their molecules and have nearly the same size.

Air is transparent and colorless. Air molecules are not blue, themselves.

At night, the sky appears transparent and colorless.

There is no "sky" on the Moon because the Moon has no atmosphere.

Our familiar blue sky requires both **air** and **sunshine**.

Procedure
PART A: RESONANCE AND SYMPATHETIC VIBRATION
Step 1: From the resonant pair of tuning forks, set one tuning fork aside.

Step 2: Using the tuning fork mallet, strike the other tuning fork and listen to the tone. A single frequency is produced by the tuning fork and amplified by the sound box it's attached to.

Step 3: Gently place your hand on the top of the tuning fork's tines to stop the vibration. Notice that the sound stops when you stop the tuning fork's vibrations.

WHY THE SKY IS BLUE

Step 4: Now arrange the two tuning forks so that the open sides of their sound boxes face each other and are 10–20 cm apart.

Step 5: Strike one tuning fork. After a few seconds, stop the struck tuning fork's vibrations by gently placing your hand on top of its tines.

What do you hear *after* the struck tuning fork is stopped?

_____

_____

The struck tuning fork's vibrations create sound waves.

Those sound waves strike the second tuning fork. The frequency of the sound waves matches the frequency of the second tuning fork.

Thus the incoming sound waves set the second tuning fork into vibration.

The second tuning fork then sends out sound waves of its own.

When the struck tuning fork is stopped, the second tuning fork continues to vibrate and emit sound.

This process is a form of **scattering:** waves from one source excite another source, and the other source sends out waves as a result.

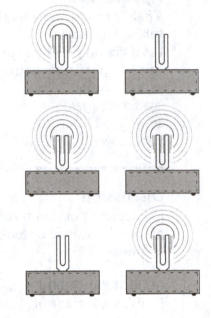

Step 6: Imagine a wide variety of tuning forks arranged in a certain region. And imagine a large number of identical tuning forks in a nearby region, as shown in Figure 1 below.

If all the tuning forks in the variety were struck, would the matching tuning forks in the other region be set into vibration? Explain.

_____

_____

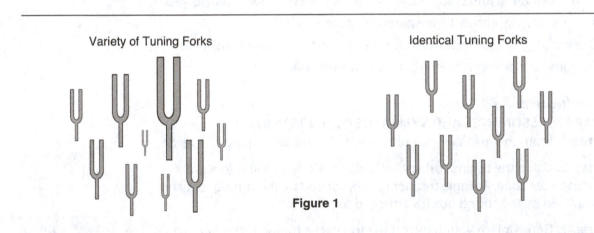

Variety of Tuning Forks

Identical Tuning Forks

Figure 1

FROM THE SUN TO THE SKY

Step 7: Answer the following questions about the Sun's light and Earth's atmosphere.

a. Does the Sun emit one frequency of electromagnetic radiation or many frequencies? Describe.

_____

_____

b. Do atmospheric molecules have a wide range of sizes or a narrow range of sizes?

_____

c. Therefore, would you expect atmospheric molecules to resonate with a wide variety of
    frequencies or with a narrow range of frequencies?

_____

d. How does the lesson of the tuning forks apply to the scattering of light in the atmosphere?

_____

_____

_____

COLOR SENSITIVITY

The spectrum of visible light runs through the colors red, orange, yellow, green, blue, and violet. Our eyes are not equally sensitive to all colors of visible light.

They are most sensitive to colors near the center of the visible spectrum (yellow/green). This color, **chartreuse,** is used for most tennis balls, many roadside worker vests, and some fire trucks.

The sensitivity of our eyes diminishes for colors at the outer extremes of the spectrum: red and violet.

ATMOSPHERIC SCATTERING

The Sun emits all frequencies of the visible spectrum and frequencies beyond the visible spectrum, such as ultraviolet and infrared.

The molecular nitrogen and oxygen in Earth's atmosphere resonate at frequencies that correspond to ultraviolet light. As a consequence, the atmosphere scatters ultraviolet light better than light of any other frequency. But a substantial amount of the Sun's ultraviolet light is blocked by ozone high in the atmosphere. And ultraviolet light is not visible.

Violet light has a frequency near that of ultraviolet, so more violet light is scattered than light of any other color in the visible spectrum. Blue is scattered to a lesser degree. Green is scattered less than blue. Yellow is scattered less than green. Orange is scattered less than yellow. And red is scattered less than orange.

WHY THE SKY IS BLUE
Step 8: Solve the puzzle.
a. Violet light is scattered more than blue light is. *Why is the sky not* **violet***?*

_____

_____

_____

b. Our eyes are more sensitive to green than they are to blue. *Why is the sky not* **green***?*

_____

_____

c. Why is the sky **blue**?

_____

_____

_____

PART B: SCATTERING IN THE TANK

Step 1: Fill the light-beam-viewing tank about three-quarters full with water.

Step 2: Gently stir enough scattering agent into the water so that it appears to have a slight visible haze to it. (This will be **much** more scattering agent than is used for other activities involving the tank.)

Step 3: Darken the room, and use the flashlight to illuminate the tank from one end, as shown in Figure 2.

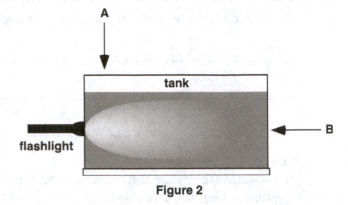

Figure 2

Step 4: Observe the tank from above. See position A shown in Figure 2. What color is observed in the light near the flashlight?

_____

Step 5: Observe the tank from the far end. See position B shown in Figure 2. What color is observed in the light that has passed through the full length of the tank?

_____

THE THICKNESS OF THE ATMOSPHERE

At midday, the Sun is at a high angle in the sky. Sunlight passes through a relatively thin layer of atmosphere before reaching the ground. In this thin layer, violet and blue are well scattered.

At sunset, sunlight passes through a thick layer of atmosphere before reaching the ground. At this greater thickness, the violet, blue, green, and yellow have been scattered.

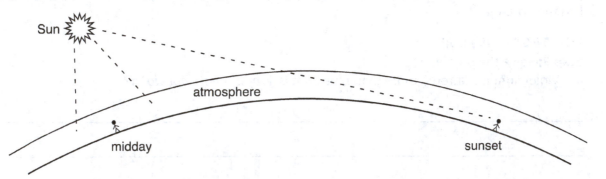

How is it that the same process can make the midday sky blue and the sunset sky red?

_____

_____

_____

CONCEPTUAL PHYSICAL SCIENCE	Activity

Light   There's a Spectrum on Your Eye

Diversion into Dispersion

Purpose

To produce a spectrum from white light and understand how it comes about

Apparatus

ray box (Rive or equivalent)
blank sheet of white paper
triangular or trapezoidal ray box prism

Discussion

One of the most famous images of Isaac Newton shows him using a glass prism to separate a narrow beam of sunlight into a spectrum of colors. This process is called **dispersion**. Pure white light consists of all the colors of the spectrum. But different colors refract to different extents. This allows the prism to separate (disperse) the colors through differential refraction. The prism is capable of a wide dispersion when the differential refraction of light upon its entering the prism is enhanced by differential refraction upon its exiting the prism. Album cover designer Storm Thorgerson created an iconic image for the cover of Pink Floyd's *Dark Side of the Moon* depicting dispersion in a prism. Lab manual author Dean Baird painted Thorgerson's design onto a 10 ft by 30 ft wall outside his classroom at Rio Americano High School in Sacramento, California.

Procedure

SETUP

Step 1: Arrange the ray box so that a single beam of white light is emitted.

Step 2: Place the sheet on the table. Place the ray box on the sheet.

Step 3: Place the prism—**dull side down**—on the paper.

CREATING A SPECTRUM

Step 1: Experiment with different arrangements (different angles of incidence on different sides of the prism) until you can produce a spectrum of colors. Your spectrum will not be wide, but you will be able to distinguish colors. Once you have it, turn up a corner of your paper to see the spectrum more clearly. Show your instructor your configuration.

If using a trapezoidal prism, use only the triangular section; there should be no light beams in the square section.

Step 2: In the space at the top of the next page, draw a **magnified** diagram of the configuration (incident ray and any refracted rays, the prism, and the emerging spectrum) as seen from directly above. Do not draw the light source in your diagram. Pay particular attention to the arrangement of the colors in the emerging beam; label red, violet, and yellow in your diagram. Include arrows to indicate the direction of travel of the beams. Use all the space; make your diagram **big!**

Step 3: Label the following on your diagram.

• **incident ray** (air to plastic)   • **internal ray** (in plastic)   • **exit ray** (with spectrum)

• **normals** (one at the air-to-plastic boundary for the first refraction, and one at the plastic-to-air boundary for the second refraction) A normal is a line perpendicular to a surface; in this case, where the incident ray hits the plastic and where the internal ray exits the plastic.

Step 4: Put your head down on the table, and arrange the apparatus so that the emerging spectrum goes into your eye directly. Move your head around so that you see different colors one at a time.

What do your partners see while you're doing this?

_____

_____

Summing Up

1. In what direction is the beam refracted as it passes from air to plastic? Mark your selection and include a diagram.
   __**toward the normal**
   __**away from the normal**
   __**not at all**

2. In what direction is the beam refracted as it passes from plastic back to air? Mark your selection and include a diagram.
   ___**toward the normal**
   ___**away from the normal**
   ___**not at all**

3. Which color undergoes the greatest amount of refraction (is bent the most in the process), and which color undergoes the least amount of refraction?

   _____

   _____

4. Which color travels the fastest in the prism (glass or plastic), and which color travels the slowest?

   _____

   _____

| **CONCEPTUAL PHYSICAL SCIENCE** | **Activity** |

Light

Pole-Arizer

Purpose
To see the production of polarized and unpolarized mechanical waves and to see how unpolarized waves become polarized

Apparatus
4 long support rods (approximately 1 meter long)
4 short crossbars (approximately 30 centimeters long)
8 right-angle clamps
1 25-ft (8-m), coiled telephone cord (or equivalent)
3 student volunteers (in addition to the instructor)

Discussion
Waves can undergo a number of processes: reflection, refraction, interference, and diffraction. Polarization is another wave phenomenon, the only one that occurs only in transverse waves. Applications of light wave polarization include 3-D movie production and the optics of sunglasses. Before exploring those applications, it's a good idea to understand the process of polarization. Because light waves cannot be observed, we will observe mechanical waves in a phone cord.

Procedure

Step 1: Prepare *two* devices like those shown in Figure 1.

a. Use two right-angle clamps to attach two crossbars to a support rod.

b. Use two more right-angle clamps to attach the two crossbars to a second support rod.

c. Repeat to make a second device.

Step 2: Ask a student volunteer to hold one end of the phone cord at chest height. The cord should be held still throughout the demonstration.

Step 3: Shake the cord vertically. Adjust the tension in the cord by changing the distance to the volunteer holder. A pulse should take approximately 1 second to get from the shaker to the holder. If the tension is too great, it's difficult to create waves with adequate amplitude. If the tension is too low, the cord will droop and hit the ground.

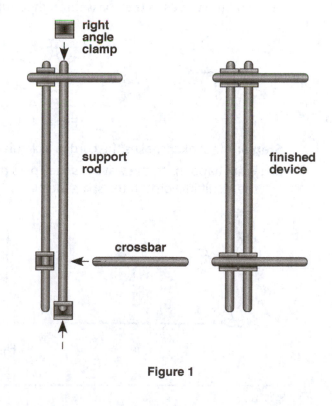

right angle clamp

support rod

crossbar

finished device

Figure 1

a. When the cord is shaken vertically, the resulting wave is **polarized**. Observers to the side would see the crests and troughs of the cord, but an observer looking down from above would see no waves.

b. When the cord is shaken in a circle, the resulting wave is **unpolarized**. Observers to the side and observers looking down from above would all see the crests and troughs.

c. What kind of wave—polarized or unpolarized—is made when the cord is shaken **horizontally?** Justify your answer.

_____

_____

Step 4: Select a second volunteer to hold one of the devices. Feed the phone cord through the gap and place the volunteer so he or she will be halfway between you and the holder. Ask the volunteer to use the handles to hold the structure vertically. See Figure 2. Position the device so that the stretched cord passes through the gap without touching very much. See Figure 3.

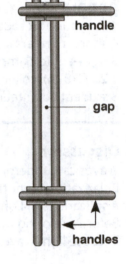

Figure 2

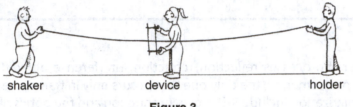

shaker   device   holder

Figure 3

Step 5: The shaker shakes the cord vertically. The device should have no effect on the waves as they pass through to the cord holder. See Figure 4.

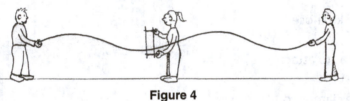

Figure 4

Step 6: The shaker shakes the cord in a circular pattern to create unpolarized waves.

a. What happens to the waves as they pass through the device? Complete Figure 5 and describe the result in words in the space below.

Figure 5

_____

_____

b. What should the device be called? See the title of the demonstration for a hint.

_____

Because the structure is made of poles, you can see how the name came about! For purposes of simplicity, we will refer to the device as a **polarizer** for the remainder of the demonstration.

Step 7: Change the orientation of the polarizer so that it is horizontal. Make sure the polarizer volunteer holds the device such in a way that his or her arms or hands don't interfere with the progress of incoming waves. What effect does the device have on unpolarized waves now?

_____

_____

Step 8: Select a **third** volunteer to hold the other polarizer. Again, feed the cord through the polarizer and arrange the volunteers so that the cord distance is divided into thirds. Have both polarizers oriented vertically to begin with. See Figure 6.

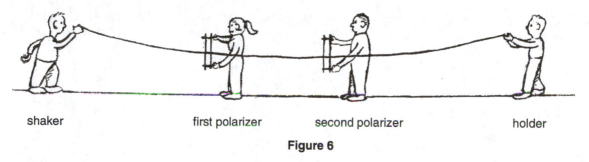

shaker   first polarizer   second polarizer   holder

Figure 6

Step 9: Shake the cord vertically. The waves should pass through both polarizers unaffected. The volunteer cord holder should receive the full energy that the shaker puts into the waves.

Step 10: Now shake the cord in a circular path to create unpolarized waves. The first polarizer will polarize the waves vertically, and the second polarizer should allow the polarized waves to pass through to the cord holder.

Step 11: While the shaker continues to make unpolarized waves and the first polarizer remains vertical, challenge the second volunteer polarizer to prevent the wave from getting through. What did he or she do to block the wave? Describe using words and pictures.

_____

_____

Step 12: If the first polarizer is oriented horizontally, the second polarizer can be used to block waves from reaching the cord holder. How can this be done?

_____

_____

Summing Up

Under what conditions can two polarizers be used to block the most energy from an unpolarized wave?

_____

_____

CONCEPTUAL PHYSICAL SCIENCE	Activity

Light   **The Polarization of Light Waves**

Blackout

Purpose

To observe the effects of the polarization of light waves, some of which are unusual or unexpected

Apparatus

bright light source such as an overhead projector or slide projector (do **not** use an LCD projector)
large sheet of polarizing film (large polarizers)
large polarizer backed with a diffuser (thin sheet of white paper or equivalent)
transparent plastic objects such as protractors, forks, and the like
small polarizing filter

Going Further

apparatus from the previous demonstration, "Pole-arizer"

Discussion

In "Pole-arizer," you saw how an unpolarized, mechanical wave could be polarized. And you saw the result of using multiple polarizers. So you know that polarized waves propagate with a preferred axis of oscillation. Light waves are electromagnetic waves in which electric and magnetic fields oscillate perpendicular to one another as the wave propagates perpendicular to those oscillations. Figure 1a. is a representation of a single, polarized ray of light. Figure 1b. shows the electric field oscillation, alone. Figure 1.c. shows a head-on view of the electric field oscillation.

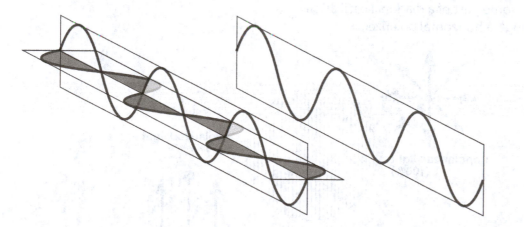

Figure 1a. The electric field oscillates vertically while the magnetic field oscillates horizontally.

Figure 1b. The electric field, alone.

Figure 1c. Head-on electric field.

Most natural light is unpolarized, meaning electric field oscillations occur in many orientations as the wave propagates forward, as shown in Figure 2a. A polarized beam is one in which a single axis of electric field oscillation is preferred, as shown in Figure 2b.

Figure 2a. Unpolarized light with many directions of electric field oscillation

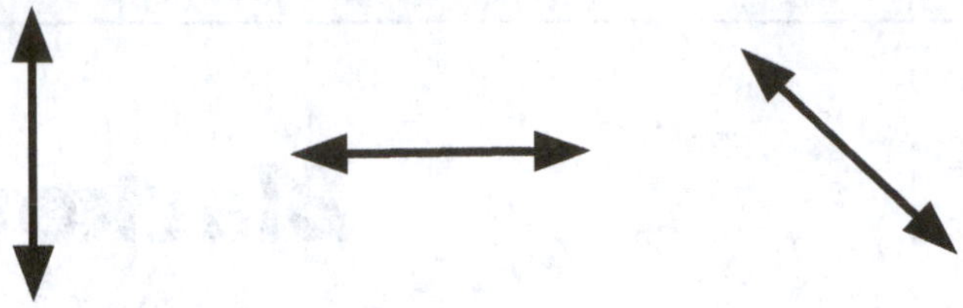

Figure 2b. Three examples of polarized light, where one direction of electric field oscillation is preferred

In this activity, you will observe the effects of polarizing light waves. You will observe crosspolarization and try to solve a polarization puzzle. You will also see optical activity in transparent plastic.

Procedure

Step 1: Shine the light beam at the diffuser-backed large polarizer as shown in Figure 3. The diffuser enables observers to see the demonstration from a wide variety of angles. The polarizer allows one axis of electric field oscillation through.

An ***ideal*** polarizer would block approximately 50% of the unpolarized light incident upon it. See Figure 4 for an illustration. Note that each diagonal oscillation has both vertical and horizontal components. So some part of a diagonal oscillation can pass through a vertical polarizer, and some part of a diagonal oscillation can pass through a horizontal polarizer.

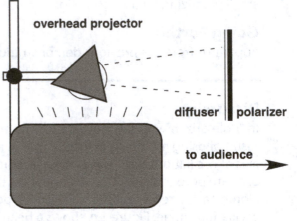

Figure 3

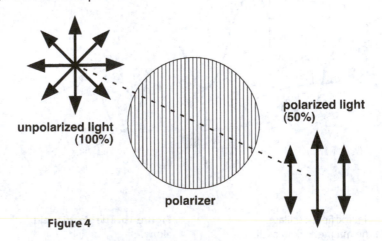

Figure 4

Observe that there is nothing peculiar about the polarized light. It may have a slight green-gray color to it. But for the most part, it appears to be nothing more than a diminished brightness of "regular," unpolarized light.

Step 2: Add a *second* polarizer (with no diffuser) in line with the first polarizer so that the optical axis (direction of polarization) matches that of the first.

Which is the better description of light passing through polarizers when their optical axes are parallel?

____Most light that passes through the first polarizer also passes through the second polarizer.
____Very little light that passes through the first polarizer also passes through the second polarizer.

Step 3: Rotate the second polarizer until its optical axis is *perpendicular* to that of the first.

Which is the better description of light passing through polarizers when their optical axes are perpendicular?

____Most light that passes through the first polarizer also passes through the second polarizer.
____Very little light that passes through the first polarizer also passes through the second polarizer.

Step 4: With the polarizers crossed, place a small polarizing filter *between* them. Orient the small filter so that its optical axis is at a 45° angle to both large polarizers. What do you see?

_____

_____

_____

_____

Consider the sequence illustrated in Figure 5 as you ponder why this happens.

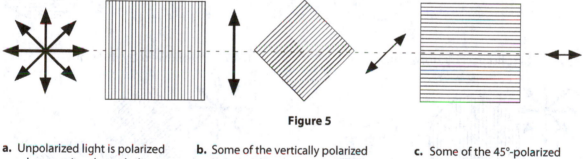

Figure 5

a. Unpolarized light is polarized when passing through the vertical filter.

b. Some of the vertically polarized light passes through the 45°-angle polarizer.

c. Some of the 45°-polarized light can pass through the horizontal filter.

Step 5: With the polarizers crossed (optical axes perpendicular to one another), place a transparent plastic object *between* them. (The object might be a protractor, a ruler, a fork, or a similar thin, solid object.) What do you see?

_____

_____

_____

Summing Up

1. Complete the illustrations of electric field oscillation below to show what happens in each polarizer configuration. Where two polarizers are used, show the oscillation **between** the filters as well as the final output.

a. One vertical filter (result is shown)

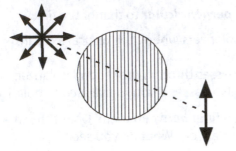

b. One horizontal filter

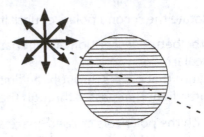

c. Vertical followed by another vertical

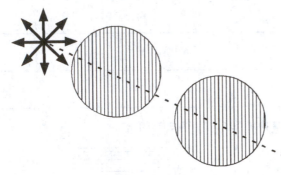

d. Horizontal followed by another horizontal

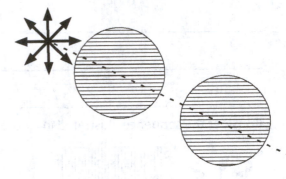

e. Vertical followed by horizontal

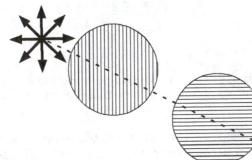

f. 45° left followed by 45° right

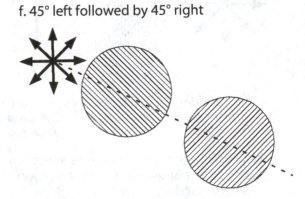

g. Vertical followed by 45° right

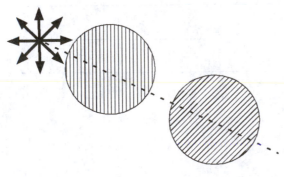

h. 45° right followed by horizontal

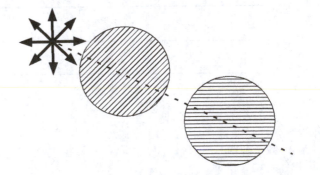

2. How might certain colors of light make it past two crossed polarizers as they did in Step 5? Describe in words and pictures.

_____

_____

_____

_____

_____

_____

_____

_____

Going Further

1. Do some research on "optical activity" to find out why the plastic objects allow polarized light to get through a perpendicular second polarizer. Record your findings on a separate sheet.

2. Can you use materials from "Pole-arizer" to simulate the demonstration in Step 4? That is, can you use a polarizer to get energy from vertically polarized waves through a horizontal polarizer? If not, why not? If so, how?

_____

_____

_____

_____

CONCEPTUAL PHYSICAL SCIENCE	Experiment

Atoms and the Periodic Table   The Size of an Atom

Thickness of a BB Pancake

Purpose
To determine the diameter of a ball bearing (BB) without directly measuring it

Apparatus
75 mL of BB shot
100-mL graduated cylinder
tray
ruler
micrometer

Discussion

This experiment distinguishes between **area** and **volume** and sets the stage for the follow-up experiment, *Oleic Acid Pancake*, where you will estimate the size of molecules. To begin, consider the diagram of the eight wooden blocks arranged to form a single $2 \times 2 \times 2$-inch cube as shown to the right. What is the surface area of this cube? There are six faces to this cube, and each face has a surface area of $2 \times 2 = 4$ square inches. The total surface area is therefore 6×4 square inches = 24 square inches.

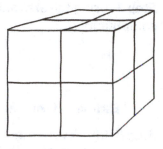

Any other arrangement of these cubes would result in a greater surface area. For example, laid out flat into a rectangular block, the surface area would be 28 square inches. If the blocks were stacked into a tower eight blocks high, the surface area would be 34 square inches. Discuss with your lab partners how the greatest surface area is obtained when the blocks are spread out as much as possible.

Although different arrangements have different surface areas, the total **volume** remains the same. Notice that as the exposed surface area of the blocks changes, the total volume of all the blocks remains the same. Similarly, the volume of pancake batter is the same whether it is in the mixing bowl or spread onto a griddle. (Except that on a **hot** griddle, the pancake gains surface area as cooking causes it to rise.)

How might you find the volume of a single pancake? The volume of a single pancake equals the surface area of one flat side multiplied by the thickness. If both the volume and the surface area are known, then the thickness can be calculated.

Volume = area × thickness

so simple rearrangement gives

$$\text{Thickness} = \frac{\text{volume}}{\text{area}}$$

Now, instead of cubical blocks or pancake batter, consider a graduated cylinder that contains BBs. The space taken up by the BBs is easily read as volume on the side of the cylinder. If the BBs are poured into a tray, their volume remains the same. Can you think of a way to estimate the diameter (or thickness) of a single BB without measuring the BB itself? Try it in this experiment and see. It will be simply another step smaller to consider the size of molecules in the next experiment.

Procedure

Step 1: Use a graduated cylinder to measure the volume of the BBs. (Note that 1 mL = 1 cm³.)

Volume = _____ cm³

Step 2: Carefully spread the BBs out to make a compact layer one pellet thick on the tray. With a ruler, determine the area covered by the BBs. Describe your procedure and show your computations.

Area = _____ cm²

Step 3: Using the area and volume of the BBs, estimate the thickness (diameter) of a BB. Show your computations.

Estimated thickness = _____ cm

Step 4: Check your estimate by using a micrometer to measure the thickness (diameter) of a BB.

Measured thickness = _____ cm

Summing Up

1. How does your estimate compare with the measurement of the diameter of the BB? Calculate the percentage error (consult Appendix C on how to do this) between the estimated thickness and the measured thickness of the BB.

_____

_____

2. Oleic acid is an organic substance that is soluble in alcohol but insoluble in water. When a drop of oleic acid is placed in water, it usually spreads out over the water surface to create a **monolayer,** a layer that is one molecule thick. From your experience with BBs, describe a method for estimating the size of an oleic acid molecule.

_____

_____

CONCEPTUAL PHYSICAL SCIENCE	Experiment

Atoms and the Periodic Table   The Size of an Atom

Oleic Acid Pancake

Purpose
To estimate the size of a single molecule of oleic acid

Apparatus
tray
water
chalk dust or lycopodium powder
eyedropper
oleic acid solution (5 mL of oleic acid in 995 mL of ethanol)
10-mL graduated cylinder

Discussion
During this experiment you will estimate the **diameter** of a single molecule of oleic acid! The procedure for measuring the diameter of a molecule will be much the same as that for measuring the diameter of a BB in the previous experiment. The diameter is calculated by dividing the volume of the drop of oleic acid by the area of the **monolayer** film that is formed. The diameter of the molecule is the depth of the monolayer.

Volume = area × depth

$$\text{Depth} = \frac{\text{volume}}{\text{area}}$$

Procedure
Step 1: Pour water into the tray to a depth of about 1 cm. So that the acid film will show itself, spread chalk dust or lycopodium powder very lightly over the surface of the water. Use a minimum amount of powder, because too much powder will prevent the oil from spreading optimally.

Step 2: Using the eyedropper, gently add a single drop of the oleic acid solution to the surface of the water. When the drop touches the water, the alcohol in it will dissolve in the water, but the oleic acid will not. The oleic acid spreads out to form a nearly circular patch on the water. Measure the diameter of the oleic acid patch in several places, and compute the average diameter of the circular patch.

    Average diameter = _____ **cm**

The average radius is, of course, half the average diameter. Now compute the area of the circle ($A = 1/2 \pi r^2$).

    Area of circle = _____ **cm²**

Step 3: Count the number of drops of oleic acid solution needed to occupy 1 mL (or 1 cm³) in the graduated cylinder. Do this three times, and find the average number of drops in 1 cm³ of solution.

>   **Number of drops in 1 cm³ =** _____

Divide 1 cm³ by the number of drops in 1 cm³ to determine the volume of a single drop.

>   **Volume of single drop =** _____ cm³

Step 4: The volume of the oleic acid alone in the circular film is much less than the volume of a single drop of the solution. The concentration of oleic acid in the solution is 5 mL per liter of solution. Every cubic centimeter of the solution thus contains only $\frac{5}{1000}$ cm³, or 0.005 cm³, of oleic acid. The volume of oleic acid in one drop is thus 0.005 of the volume of one drop. Multiply the volume of a drop by 0.005 to find the volume of oleic acid in the drop. This is the volume of the layer of acid in the tray.

>   **Volume of oleic acid =** _____ cm³

Step 5: Estimate the diameter of an oleic acid molecule by dividing the volume of oleic acid by the area of the circle.

>   **Diameter =** _____ cm

The diameter of an oleic acid molecule as obtained by this method is good, but not precise. This is because an oleic acid molecule is not spherical but, rather, elongated like a hot dog. One end is attracted to water, and the other end points away from the water surface. The molecules stand up like people in a puddle! So the estimated diameter is actually the estimated length of the short side of an oleic acid molecule.

Summing Up

1. What is meant by the term ***monolayer***?

_____

2. Why is it necessary to dilute the oleic acid for this experiment? Why alcohol?

_____

3. The shape of an oleic acid molecule is more like that of a hot dog than like a sphere. Furthermore, one end is attracted to water (***hydrophilic***), so the molecule stands up on the surface of water. Assume an oleic molecule is 10 times longer than it is wide. Then estimate the volume of one oleic acid molecule.

_____

_____

CONCEPTUAL PHYSICAL SCIENCE   **Experiment**

Atoms and the Periodic Table   Spectroscopy

Bright Lights

Purpose

To examine light emitted by various elements when they are heated, and to identify a component in an unknown salt by examining the light emitted by the heated salt

Apparatus

EQUIPMENT
Bunsen burner
heatproof glove or potholder
metal spatulas or flame test wires (nichrome)
diffraction gratings
spectroscope (commercial or homemade)
ring stand with clamp large enough to hold spectroscope
colored pencils
discharge tubes containing various gases

CHEMICALS
0.1 M HCl solution
sodium chloride powder
potassium chloride powder
calcium chloride powder
strontium chloride powder
barium chloride powder
cupric chloride powder

Discussion

When materials are heated, their elements emit light. The color of the light is characteristic of the types of elements in the heated material. Strontium, for example, emits predominantly red light, and copper emits predominantly green light. An element emits light when its electrons make a transition from a higher energy level to a lower energy level. Every element has its own characteristic pattern of energy levels and, therefore, emits its own characteristic pattern of light frequencies (colors) when heated.

It is interesting to look at the light emitted by elements through either a diffraction grating or a prism. Rather than producing all colors, the elements produce a spectrum showing only particular colors (frequencies). When the emitted light passes first through a thin slit, and then through the grating or prism, the different colors appear as a series of vertical lines, as shown in Figure 1. (The vertical lines are images of the original slit.) Each vertical line corresponds to a particular energy transition for an electron in an atom of the heated element. The pattern of lines, called an **emission spectrum,** is characteristic of the element. It is often used as an identifying feature—much like a fingerprint. Astronomers, for instance, can tell the elemental composition of stars by examining their emission spectra.

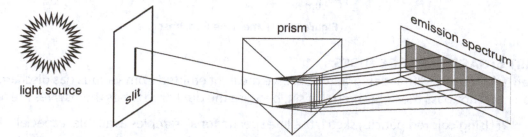

Figure 1. When heated, a gaseous element produces a discontinuous emission spectrum.

Procedure

PART A: FLAME TESTS

Step 1: Using either a heatproof glove or a potholder, hold the tip of a metal spatula (or nichrome wire, if available) in a Bunsen burner flame until the spatula tip is red hot, and then dip it in a 0.1 M HCl solution. Repeat this cleaning process several times until you no longer see color coming from the metal when it is heated.

Step 2: Obtain small amounts of the metal salts to be tested, and label each sample. Dip the spatula tip first into the HCl solution and then into one of the salts, so that the tip becomes coated with the powder. Then put the tip into the flame and observe the color. Record your observations on the report sheet.

Step 3: Rinse the spatula in water, and then clean as described in Step 1.

Step 4: Repeat this procedure for all the salts, being sure to clean the spatula each time.

PART B: FLAME TESTS USING A DIFFRACTION GRATING

Step 1: Look at the lights in your classroom through a diffraction grating. Note how the diffraction grating separates the white light into a rainbow of all colors.

Step 2: After your instructor darkens the room, repeat the procedure of Part A. This time, however, observe the flame through your diffraction grating. Note how the flame produced by the salt produces only a select number of colors.

Step 3: To help you distinguish which colors the salts are emitting, use a spectroscope. You can use either a commercial spectroscope or a homemade one as shown in Figure 2.

Step 4: On your report sheet, sketch the predominant lines you observe for each salt, using colored pencils. You will see lines both to the left and to the right of the slit. Sketch only the lines to the right. (*Note*: Some salts will also show regions of continuous color.)

Step 5: Obtain an unknown metal salt from your instructor, and record its number.

Step 6: Observe and sketch the line spectrum for your unknown salt, using colored pencils.

Step 7: Identify your unknown salt, based on its spectrum.

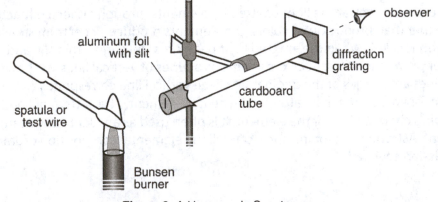

Figure 2. A Homemade Spectroscope

PART C: GAS DISCHARGE TUBES

Step 1: Through a diffraction grating, observe the light emitted from various gas discharge tubes. (You need not pass the light through a slit, because the discharge tubes themselves are narrow.)

Step 2: Using colored pencils, sketch the line spectra for all samples available, especially hydrogen, oxygen, and water vapor.

Bright Lights Report Sheet

PART A: FLAME TEST

Compound	NaCl	KCl	$CaCl_2$	$SrCl_2$	$BaCl_2$	$CuCl_2$
Color						

PART B: FLAME TESTS USING A SPECTROSCOPE

NaCl	Sketch of line spectrum
	V I B G Y O R

$SrCl_2$	Sketch of line spectrum
	V I B G Y O R

KCl	Sketch of line spectrum
	V I B G Y O R

$BaCl_2$	Sketch of line spectrum
	V I B G Y O R

$CaCl_2$	Sketch of line spectrum
	V I B G Y O R

$CuCl_2$	Sketch of line spectrum
	V I B G Y O R

Unknown number _____

Unknown	Sketch of line spectrum
	V I B G Y O R

Identity of unknown: _____

PART C: DISCHARGE TUBES

Substance	Sketch of line spectrum
	V I B G Y O R

Substance	Sketch of line spectrum
	V I B G Y O R

Substance	Sketch of line spectrum
	V I B G Y O R

Substance	Sketch of line spectrum
	V I B G Y O R

Substance	Sketch of line spectrum
	V I B G Y O R

Substance	Sketch of line spectrum
	V I B G Y O R

Substance	Sketch of line spectrum
	V I B G Y O R

Substance	Sketch of line spectrum
	V I B G Y O R

Summing Up

1. When the spatula was initially being cleaned in the flame, it may have given off yellow light. If this happened, what residue was probably on the spatula before it was cleaned?

_____

_____

2. What produces the colors of fireworks?

_____

_____

3. Is the gas in a blue "neon lamp" actually neon? Explain.

_____

_____

4. Does the line spectrum of water vapor bear any resemblance to the line spectra of hydrogen and oxygen? Why or why not?

_____

_____

CONCEPTUAL PHYSICAL SCIENCE	Activity

The Atomic Nucleus and Radioactivity   Radioactive Half-Life

Get a Half-Life

Purpose
To simulate radioactive decay half-life

Apparatus
25 playing dice (regular, cube)

Discussion
The rate of decay for a radioactive isotope is measured in terms of half-life—the time for one-half of a radioactive quantity to decay. Each radioactive isotope has its own characteristic half-life (Table 1). For example, the naturally occurring isotope of uranium, uranium-238, decays into thorium-234 with a half-life of 4.5 billion years. This means that only half of an original amount of uranium-238 remains after this time. After another 4.5 billion years, half of *this* decays, leaving only one-fourth of the original amount remaining. Compare this with the decay of polonium-214, which has a half-life of 0.00016 second. With such a short half-life, any sample of polonium-214 will quickly disintegrate.

Table 1

Isotope	Half-Life
Uranium-238	4,500,000,000 years
Plutonium-239	24,400 years
Carbon-14	5,730 years
Lead-210	20.4 years
Bismuth-210	5.0 days
Polonium-214	0.00016 second

The half-life of an isotope can be calculated by the amount of radiation coming from a known quantity. In general, the shorter the half-life of a substance, the faster it decays, and the more radioactivity per amount of the substance is detected.

In this activity, you will investigate three hypothetical substances, each represented by a designation on the face of a die. The first substance, represented by a given symbol, is marked on only one side of the die. The second substance, represented by two symbols, is marked on two sides of the die, and the third substance, represented by three symbols, is marked on three sides of the die. Rolling a large number of these identical dice simulates the process of decay for these substances. As a substance's symbol turns face up, it is considered to have decayed and is removed from the pile. This process is repeated until all of the dice have been removed. Since the symbol of the first substance is only on one side, this substance will decay the slowest (because its symbol will fall face up least frequently, and it will stay in the game longer). The second substance, marked on two sides, will decay faster, requiring fewer rolls before all the dice are removed. The third substance, marked on three sides, will decay the fastest. After tabulating and graphing the numbers of dice that decay in each roll for these simulated substances, you will be able to determine their half-lives.

Procedure

Step 1: Shake the cubes in a container, and roll them onto a flat surface. When rolled, each die can come up as a one, two, three, four, five, or six.

Step 2: Count the dice that came up as ones, and record this number under "Removed" in the data table.

Step 3: Remove the ones in a pile off to the side.

Step 4: Gather the remaining dice back into the container, and roll them again.

Step 5: Repeat Steps 2–4 until all dice have been counted, tabulated, and set aside.

Step 6: Repeat Steps 1–5, removing dice that come up as threes or fives (odds greater than one).

Step 7: Repeat Steps 1–5, removing dice that come up even (twos, fours, or sixes).

Data Table

Throw	First Substance Ones (·)		Second Substance Non-One Odds (· · · & : · :)		Third Substance Evens (· · , : : , and : : :)	
	Removed	Remaining	Removed	Remaining	Removed	Remaining
Initial Count						
1						
2						
3						
4						
5						
6						
7						
8						
9						
10						
11						
12						
13						
14						
15						
16						
17						
18						
19						
20						
21						
22						
23						
24						
25						

Step 8: Plot the number of dice (cubes) remaining vs. the number of throws for each substance on the following graph. Use a different color or line pattern to graph the results for each substance. For each substance, draw a single smooth line or curve that approximately connects all points. **Do not connect the dots!** Indicate your color or line pattern code below the graph.

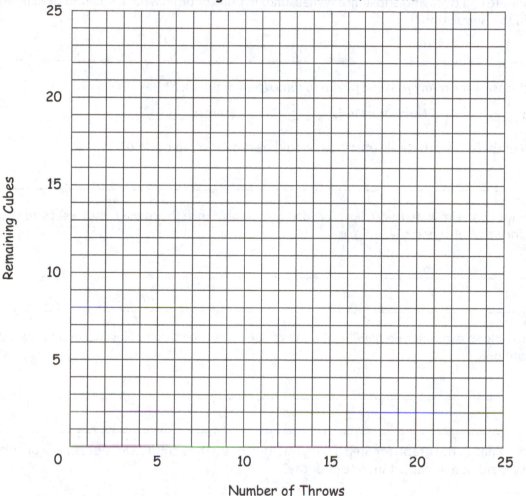

Remaining Cubes vs. Number of Throws

Remaining Cubes (y-axis)

Number of Throws (x-axis)

First substance (one side on die) line color or pattern:

Second substance (two sides on die) line color or pattern:

Third substance (three sides on die) line color or pattern:

Summing Up

1. How many rolls did it take for the number of each dice set to be reduced by half? These are your half-life readings.

   Ones: _____   **Non-One Odds:** _____   **Evens:** _____

2. The half-life of a decaying substance is measured in units of time. What is the unit of half-life used in this simulation?

   _____

3. In each case, how many rolls did it take to remove all of the dice?

   Ones: _____   **Non-One Odds:** _____   **Evens:** _____

4. Which of these hypothetical substances would be the most radioactive?

   _____

5. How might you simulate the radioactive decay of a substance that decays into a second substance that also decays?

   _____

   _____

6. Is it possible to estimate the half-life of a substance in a single throw? How accurate might this estimate be?

   _____

   _____

7. Are your lines in the graph for Step 8 fairly straight, or do they curve? Do these lines correspond to a constant or a nonconstant rate of decay?

   _____

   _____

8. a. Substance X has a half-life of 10 years. If you start with 1000 g, how much will be left after:

   i.   10 years? _____

   ii. 20 years? _____

   iii. 50 years? _____

   iv. 100 years? _____

   b. Will this sample of substance X ever totally disappear? If so, estimate how soon. If not, explain.

   _____

   _____

CONCEPTUAL PHYSICAL SCIENCE	Activity

The Atomic Nucleus and Radioactivity   **Radiometric Dating Simulation**

Radioactive Speed Dating

Purpose
To correctly estimate the age of various virtual objects (skulls, rocks, and the like) using the principles of radiometric dating

Apparatus
computer
PhET simulation "Radioactive Dating Game" (available at http://phet.colorado.edu)

Discussion
Radiometric dating is a nuclear-decay-based method for determining the age of very old things. Here's how it works: Radioactive nuclei transform from one nuclear structure to another at a mathematically reliable rate. In a given amount of time, half the original nuclei have decayed to the new nuclear structure. That given amount of time is called the "half-life." After one half-life, half of the original nucleus remains. After two half-lives, a quarter remains, and so on. By measuring the percent of the original nuclei that remain, it is possible to determine the age of the decaying sample. For example, carbon-14 has a half-life of 5700 years. If an object were found in which 25.0% of the original carbon-14 remained, the age of the object would be taken as approximately 11,400 years. Other percentages can be translated to ages using an exponential function.

Procedure
Step 1: Open the "Radioactive Dating Game" simulation. Click on the "Dating Game" tab in the simulation window.

Step 2: Use scientific techniques applicable to the simulation to estimate the ages of all the objects that allow estimates. Do not attempt to hack. Do not use any other reference sources.
Do not cheat.

Step 3: When you have entered six correct estimates in Table 1, record the time displayed on the computer screen (this is not a part of the sim). Obtain a stamp from the instructor. Reward credit begins at six "smilies" and increases with each additional three "smilies." Record the time, and obtain a stamp in Table 1 at each milestone.

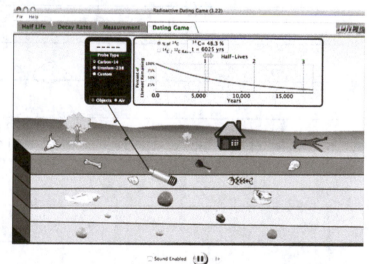

Figure 1. Radioactive Dating Game Sim in the Dating Game Tab

Table 1

"Smilies"	Total Reward	Computer Time	Instructor's Stamp
6			
9			
12			
15			
18			

Summing Up

1. Carbon dating isn't a valid technique for dating fossils beyond a certain age. And it's not a valid technique for dating inorganic items (such as rocks). How did you overcome its limitations?

_____

_____

_____

_____

2. What are the limitations of uranium-238 dating? How did you overcome its limitations?

_____

_____

_____

_____

3. Two "accepted" ages are clearly in error. What are they, and why do they produce invalid "correct" responses? (Scientists do not use radiometric dating techniques on such objects.)

_____

_____

_____

_____

CONCEPTUAL PHYSICAL SCIENCE	Activity

The Atomic Nucleus and Radioactivity   **Nuclear Fission**

Chain Reaction

Purpose
To simulate a simple chain reaction

Apparatus
100 dominoes
large table or floor space
stopwatch

Discussion
Give your cold to two people, who in turn give it to two others, who in turn do the same on down the line, and before you know it, everyone in class is sneezing. You have set off a chain reaction. This is similar to what happens within a lump of uranium-235 where one neutron triggers the release of two or more neutrons, which trigger the release of even more neutrons. That one neutron has set off a chain reaction. Very quickly, so many neutrons are produced that the lump of uranium explodes as an atomic bomb.

In this activity, you'll explore this idea of the chain reaction.

Procedure
Step 1: Set up a strand of dominoes about half a domino length apart in a straight line. Gently push the first domino over, and measure how long it takes for the entire strand to fall over.

Step 2: Arrange the dominoes again, this time as shown in the figure, so that when one domino falls, another one *or two* are toppled. These topple others in chain-reaction fashion. Set up until you run out of dominoes or table space. When you finish, push the first domino over and measure how long it takes for all the dominoes to fall. Roughly compare the number of falling dominoes per second at the beginning with the number falling per second at the end.

Summing Up
1. Which reaction, wide-spaced or close-spaced dominoes, took a shorter time?

    _____

    _____

2. How did the number of dominoes being knocked over per second change for each reaction?

    _____

    _____

3. What caused each reaction to stop?

_____

_____

4. Now imagine that the dominoes are the neutrons released by uranium atoms when they undergo **fission** (split apart). Neutrons from the nucleus of a fissioning uranium atom hit other uranium atoms and cause *them* to undergo fission. This reaction continues to grow if there are no controls. Such an uncontrolled reaction occurs in a split second and is called a **nuclear explosion.** How is the domino chain reaction similar to the nuclear fission process?

_____

_____

_____

_____

5. How is the domino reaction dissimilar to the nuclear fission process?

_____

_____

_____

_____

CONCEPTUAL PHYSICAL SCIENCE	Activity

Elements of Chemistry   Physical and Chemical Properties

Chemical Personalities

Purpose
To identify various physical and chemical properties of matter and distinguish between physical and chemical changes

Apparatus

EQUIPMENT
thermometer with cork holder
ring stand with two clamps
hot plate
250-mL and 500-mL beakers
large test tube
boiling chips
glass stirring rod
well-plates (ceramic or glass)
eyedroppers
microspatula
evaporating dishes
ice

CHEMICALS
various elements and compounds
methanol
iodine crystals
sucrose crystals
acetone
steel wool
cupric sulfate pentahydrate crystals
10% sodium carbonate solution
10% sodium sulfate solution
1 M HCl
10% sodium chloride solution
10% calcium chloride solution
white distilled vinegar

Discussion
Many physical properties can be observed using our senses; color, crystal shape, and phase at room temperature are some examples. Other physical properties involve quantitative observations and so must be measured; density and boiling point are two examples. A *physical change* is any change in a substance that does not involve a change in its chemical composition. During a physical change, no new chemical bonds are formed, so the chemical composition remains the same. Examples of physical change include boiling, freezing, expanding, and dissolving.

Matter can also be characterized by its chemical properties. The chemical properties of a substance include all the chemical changes possible for that substance. For example, it is a chemical property of gasoline to burn when ignited in the presence of oxygen. A *chemical change* is one in which the substance is transformed to a new substance. That is, there is a change in the chemical composition of the substance. During a chemical change, the atoms are pulled apart from one another, rearranged, and put back in a new arrangement. Examples of chemical change include burning, rusting, fermenting, and decomposing.

In this experiment, you will first identify and record various physical properties of substances, using qualitative observations, such as changes in color or phase, and quantitative observations, such as boiling points. In the second part, you will look at changes in matter and determine whether they are physical or chemical.

Procedure

PART A: PHYSICAL PROPERTIES

Step 1: Examine the various substances provided by your instructor, and record your observations in Table 1 of the report sheet. (*Note*: Some substances may be toxic. As a precaution, do not open any containers without the permission of your instructor.)

Step 2: Assemble the apparatus shown in Figure 1. Add about 300 mL of water to the 500-mL beaker and about 3 mL of methanol, CH_3OH, to the test tube. Do not forget to add a boiling chip to the test tube. Turn on the hot plate to medium. Use a stirring rod to stir the water while it is being heated, and pay close attention to the thermometer readings. Note that as the methanol boils, its temperature remains constant. Record the boiling point.

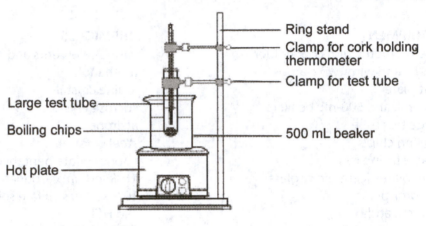

Labels:
- Ring stand
- Clamp for cork holding thermometer
- Clamp for test tube
- Large test tube
- Boiling chips
- 500 mL beaker
- Hot plate

Figure 1

Step 3a: Place a small crystal of iodine in a well of a ceramic or glass well-plate, and place a small crystal of sucrose in a second well-plate. Use an eyedropper to fill each well with distilled water, and stir gently with a microspatula. Record whether each substance is completely soluble, partially soluble, or insoluble. Rinse the iodine into a designated waste container and the sucrose into the sink.

Step 3b: Repeat the procedure using acetone as the solvent. To clean up, you may wipe any remaining iodine on the well-plate with a paper towel wetted with acetone.

PART B: PHYSICAL AND CHEMICAL CHANGES

Complete Table 2 of the report sheet for each of the following systems.

Step 1: Inspect a small piece of steel wool, and wet it with vinegar. Shake off any residual vinegar. Place the steel wool in a ceramic dish, and briefly heat on a hot plate set to high. Allow the system to cool to room temperature. Observe and record any changes in the steel wool.

Step 2: Inspect some cupric sulfate pentahydrate crystals, $CuSO_4 \cdot 5H_2O$. Place a few crystals in a ceramic dish, and heat on a hot plate set to medium. Observe and record any changes in the salt. After the system has cooled to room temperature, add a few drops of water to the crystals. Observe and record any changes.

Step 3: Place a few drops of a 10% sodium carbonate solution, Na_2CO_3, in one well of a well-plate and a few drops of a 10% sodium sulfate solution, Na_2SO_4, in a second well of the same well-plate. Add two or three drops of 1 M hydrochloric acid to each well. Observe and record any changes.

Step 4: Place a few drops of a 10% sodium chloride solution, NaCl, in one well of a well-plate and a few drops of a 10% calcium chloride solution, $CaCl_2$, in a second well of the same well-plate. Add several drops of a 10% sodium carbonate solution to each well. Observe and record any changes.

Step 5: Inspect some iodine crystals, I_2. Place a few of the crystals in a dry 250-mL beaker, and cover with an evaporating dish that contains ice, as shown in Figure 2. *In a fume hood*, place the beaker on a hot plate set to medium. Observe and record any changes.

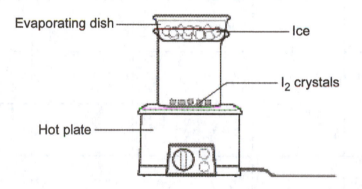

Figure 2

CHEMICAL PERSONALITIES

Name _____ Section _____ Date _____

Physical and Chemical Properties and Changes Report Sheet

PART A: PHYSICAL PROPERTIES
1. Complete Table 1.

Table 1

Name of Substance	Chemical Formula	Phase at Room Temperature	Color	Other Physical Properties Observed	Element or Compound?

2. Boiling point of methanol _____ °C

3. Solubility (Record "I" for Insoluble, "PS" for Partially Soluble, or "S" for Soluble.)

      Iodine in water _____   Sucrose in water _____

      Iodine in acetone _____   Sucrose in acetone _____

PART B: PHYSICAL AND CHEMICAL CHANGES
1. Complete Table 2.

Table 2

Procedure	Observation	Physical Change or Chemical Change?	Evidence or Reasoning
1. steel wool + heat			
2a. $CuSO_4 \cdot 5H_2O$ + heat			
2b. $CuSO_4 + H_2O$			
3a. $Na_2CO_3 + HCl$			
3b. $Na_2SO_4 + HCl$			
4a. $NaCl + Na_2CO_3$			
4b. $CaCl_2 + Na_2CO_3$			
5. I_2 + heat			

Name _____ Section _____ Date _____

1. Distinguish between a qualitative observation and a quantitative observation. Give an example of each from this experiment.

2. Classify the following properties of sodium metal as physical or chemical:
   a. silver metallic color _____
   b. turns gray in air _____
   c. melts at 98°C _____
   d. reacts explosively with chlorine _____

3. Classify the following changes as physical or chemical:
   a. steam condenses to liquid water on a cool surface _____
   b. baking soda dissolves in vinegar, producing bubbles _____
   c. mothballs gradually disappear at room temperature_____
   d. baking soda loses mass as it is heated _____

CONCEPTUAL PHYSICAL SCIENCE	Experiment

Elements of Chemistry   Qualitative Analysis

Mystery Powders

Purpose
To identify nine common household chemicals by using qualitative analysis, a method of determining a substance's identity by subjecting it to a series of tests

Apparatus

HOUSEHOLD CHEMICALS
cornstarch, $(C_6H_{10}O_5)n$
white chalk, $CaCO_3$
Plaster of Paris, $2CaSO_4 \cdot H_2O$
washing soda, Na_2CO_3
Epsom salt, $MgSO_4 \cdot 7H_2O$
baking soda, $NaHCO_3$
boric acid, H_3BO_3
table sugar, $C_{12}H_{22}O_{11}$
table salt, $NaCl$

TEST REAGENTS
tincture of iodine
phenolphthalein
white vinegar
rubbing (isopropyl) alcohol (70%)
sodium hydroxide (0.3M)

EQUIPMENT
test tubes*
eye dropper
safety goggles
beaker
spatula

Note to instructor: for easier clean-up, glass or plastic well plates may be substituted for test tubes.

Safety
You will be working with unknown chemicals. Handle them carefully and never, ever taste them. Some may taste sweet, but others may burn your tongue! If you are unsure about any procedure, ask your instructor. Avoid spillage. Use small quantities—no more than is required for each test. Never place unused unknown back in the reagent bottle, because this can contaminate the stock material. Consult your instructor for proper disposal.

Discussion
You are given nine vials, and each contains a white powder that is a common household chemical. Your task is to identify these unknowns on the basis of their different physical and chemical properties. For this experiment you should develop a qualitative analysis scheme, such as shown in Figure 1, to show how the chemicals can be systematically identified.

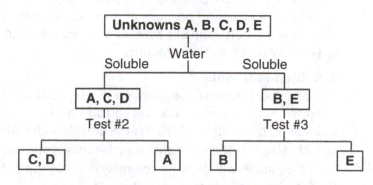

Figure 1. A Sample Qualitative Analysis Scheme

Procedure

Using the tests outlined below, prepare a qualitative analysis scheme that should enable you to unequivocally determine the identity of each unknown. This scheme should be prepared before conducting this experiment. It is recommended that you start with Test 1 (Solubility in Water). You may then choose your own order of testing. Some orders are more efficient than others. Try to develop a scheme that minimizes the number of tests that must be conducted. Use Table 1 as a guide to the physical and chemical properties of the household chemicals to be identified.

Tests

1. Solubility in Water

Place a pea-sized portion of the unknown solid in a test tube, and add about 5 mL of water. To mix the contents, hold the top of the test tube securely between your thumb and index finger, and gently slap the bottom of the test tube with the index finger of your opposite hand. Be careful so that none of the solution spills out. Consult Table 1 for behaviors of each of the household chemicals.

2. Tincture of Iodine

This test may be used for any unknown that is insoluble in water. Add a few drops of tincture of iodine to the unknown as it sits undissolved in water. A deep blue color forms as the iodine complexes with cornstarch. A brownish color will appear for all other unknowns insoluble in water.

3. Phenolphthalein

This test may be used for any unknown that is soluble in water. Add a drop of phenolphthalein to the dissolved unknown, and a bright pink color will result if the solution is very alkaline. This test is positive for washing soda, Na_2CO_3, which turns a medium pink, and it is also positive for baking soda, $NaHCO_3$, which turns a light pink.

4. White Vinegar

This test is applicable to all unknowns. Add a few drops of vinegar to the unknown, in either a dissolved or an undissolved state. The formation of bubbles is a sign of the carbonate ion [CO_3^{2-}], which decomposes to gaseous carbon dioxide upon treatment with an acid (vinegar). This test is positive for white chalk, $CaCO_3$; washing soda, Na_2CO_3; and baking soda, $NaHCO_3$.

5. Sodium Hydroxide (0.3M)

This is a specific test for magnesium sulfate, $MgSO_4$, which is found in Epsom salt. If the unknown dissolves in water and forms an insoluble precipitate when treated with this test reagent, then the unknown contains magnesium sulfate.

6. Hot Water

All the water-soluble unknowns become markedly more soluble in warmer water, with the exception of sodium chloride, NaCl (table salt). This test, therefore, is specific to sodium chloride. Place several pea-sized portions of the unknown solid in a test tube, along with 5 mL of water. Heat in a hot water bath held at about 60°C. No marked improvement in solubility suggests that the unknown may be sodium chloride.

7. Rubbing Alcohol

This test may be used for any unknown that is soluble in water. Place a pea-sized portion of the unknown solid in a test tube, and add about 7 mL of rubbing alcohol (isopropyl, 70%). Of the unknowns, boric acid, H_3BO_3, has the greatest solubility in rubbing alcohol. The solubility of Epsom salt, $MgSO_4$, in rubbing alcohol is very minimal and might not be readily observed. The following four substances will show no solubility in rubbing alcohol: baking soda, $NaHCO_3$; washing soda, Na_2CO_3; sugar, $C_{12}H_{22}O_{11}$; and salt, NaCl.

Table 1. Properties of nine household chemicals

Household Chemical	Reactions to Tests						
	1 Solubility	2 Iodine	3 pH > 7	4 Vinegar	5 Hydroxide	6 Hot water	7 Alcohol
Cornstarch	insoluble	positive	—	negative	—	—	—
White chalk	insoluble	negative	—	positive*	—	—	—
Plaster of Paris	insoluble	negative	—	negative	—	—	—
Washing soda	soluble	—	medium pink	positive	negative	greater sol.	insoluble
Epsom salt	soluble	—	negative	negative	positive	greater sol.	mostly insoluble
Baking soda	soluble	—	light pink	positive	negative	greater sol.	insoluble
Boric acid	moderately soluble	—	negative	negative	negative	slight increase	soluble
Table sugar	soluble	—	negative	negative	negative	greater sol.	insoluble
Table salt	soluble	—	negative	negative	negative	no effect	insoluble

* Some brands of chalk are not made of calcium carbonate, and their results on the vinegar test are negative.

Summing Up

1. Identify your unknowns.

    Corn starch was unknown: _____   **White chalk was unknown:** _____

    Plaster of Paris was unknown: _____   **Washing soda was unknown:** _____

    Epsom salt was unknown: _____   **Baking soda was unknown:** _____

    Boric acid was unknown: _____   **Table sugar was unknown:** _____

    Table sugar was unknown: _____

2. Which of the tests used in this experiment measure physical properties, and which measure chemical properties?

    _____

3. Sugar is a chemical, but it is also a food. Is this contradictory? Briefly explain.

    _____

4. How many individual tests did you have to perform to identify all of the unknowns? (Review your qualitative analysis scheme, or simply count your dirty test tubes.) Compare your number of tests with those of your classmates.

    _____

MYSTERY POWDERS

CONCEPTUAL PHYSICAL SCIENCE	Activity

How Atoms Bond and Molecules Attract   **Electron Dot Diagrams**

Dot to Dot

Purpose
To practice writing plausible electron dot structures for simple molecules

Apparatus
periodic table

Discussion
To help predict how atoms bond together, you can use the *octet rule*, which states that atoms form chemical bonds so as to have a filled, outermost occupied shell. Looking at a periodic table, we see that the noble gases already have filled the outermost occupied shells, which explains why they tend not to form chemical bonds. Note that all the noble gases except helium have *eight* outermost electrons—hence the name "octet rule."

The octet rule can be used to build an ammonia molecule, NH_3. Knowing that a nitrogen atom has five valence electrons and a hydrogen atom has one, you can satisfy the octet rule with the electron dot structure

$$H : \overset{\displaystyle ..}{\underset{\displaystyle ..}{N}} : H$$
$$H$$

(eight election surrounding N, two electrons surrounding H)

In this structure, the nitrogen atom has eight valence electrons so that its outermost occupied shell, which has a capacity of eight electrons, is filled. Each hydrogen atom has two valence electrons so that its outermost occupied shell, which has a capacity of two electrons, is also filled.

Using the octet rule, you will learn to draw electron dot structures of simple covalent compounds, given their chemical formulas. (This activity focuses on bonding in simple covalent molecules that obey the octet rule. There are many molecules that do not follow this rule, but they are not introduced in this laboratory.)

To construct a plausible electron dot structure for a molecule:

1. Determine the total number of valence electrons available from the chemical formula by adding up the valence electrons of all atoms in the molecule.

2. Write the chemical symbols for all atoms in the molecule, arranging the symbols as you think they might appear in the molecule. Many molecules contain a central atom. If there are many elements present, place them in the order in which they are written in the formula.

3. Place single bonds between all pairs of atoms, remembering that each bond represents *two* electrons, one from each atom.

4. Add the appropriate number of electrons around each atom so that the atom obeys the octet rule (eight electrons around all atoms other than hydrogen, two around hydrogen).

5. Count the number of electrons used in your structure. If the number is equal to the sum you calculated in Step 1, your structure is plausible and you are done. If the number of electrons used is greater than the sum you calculated in Step 1, try putting in multiple bonds (try double first, then triple) until all atoms obey the octet rule and the number of electrons used matches the number calculated in Step 1.

6. If the number of electrons used is less than the sum you calculated in Step 2, you probably made an error in your count in Step 1, so you should recalculate.

Procedure

PART A: DRAWING ELECTRON DOT STRUCTURES

To show these steps in action, here are examples using carbon tetrabromide and carbon monoxide.

1. Total number of valence electrons available:

Carbon tetrabromide, CBr_4
1 C atom (4 valence electrons each) $1 \times 4 =$ 4
4 Br atoms (7 valence electrons each) $4 \times 7 = \underline{28}$
Total   32

Carbon monoxide, CO
1 C atom (4 valence electrons) $1 \times 4 =$ 4
1 O atom (6 valence electrons) $1 \times 6 = \underline{6}$
Total   10

2. Arrangement of symbols for all atoms (locate central atom if applicable):

Br

Br C Br

Br

C O

3. Single bonds between all pairs of atoms:

C : O

4. Add remaining electrons to complete octets:

5. Count the number of electrons used:

The above structure uses 32 electrons, which is equal to the number calculated in Step 1. The structure is complete.

The above structure uses 14 electrons, more than the total number calculated in Step 1. A multiple bond must be present. With a double bond, the structure would be

$$:C::O:$$

There is still a problem because the structure uses 12 electrons. A triple bond must be present:

$$:C:::O:$$

Finally, the structure obeys the octet rule and uses the total number of valence electrons available. The structure is complete.

Many chemists prefer to use a line (or lines), rather than dots, to represent a bonded pair of electrons. With this line notation, these two molecules are represented as

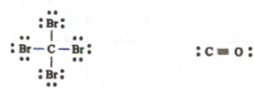

PART B: ELECTRON DOT STRUCTURES REPORT SHEET

Complete the following table.

Chemical Formula	Total Number of Valence Electrons	Plausible Electron Dot Structure
HCl		
SiH_4		
PH_3		
O_3		
HOCl		
HCN		
CO_2		
PCl_3		
CH_2Cl_2		
H_2O_2		
H_2CO_2		

CONCEPTUAL PHYSICAL SCIENCE	Activity

How Atoms Bond and Molecules Attract   VSEPR Theory

Repulsive Dots

Purpose
To become familiar with the three-dimensional shapes of molecules, build molecular models from information given in electron dot structures, draw electron dot structures from information given in a molecular model, and predict the polarity of a molecule from its molecular shape

Apparatus
molecular model kit
built models of six different molecules

Discussion
It is extremely useful for a chemist to know the three-dimensional shape of the molecules of a given compound. For some compounds (those containing two or more polar bonds), knowledge of molecular shape is necessary for predicting polarity. One can illustrate the relationship between molecular shape and polarity using water as an example. The O—H bond is polar because of the difference in the electronegativities of the two elements. A water molecule contains two O—H bonds. If a water molecule were linear, as shown in Figure 1, the two polar bonds would be equal in magnitude, but opposite in direction. Their effects would cancel, and the water molecule as a whole would be nonpolar. However, water molecules have a bent shape, as shown in Figure 2. In such a structure, the two polar bonds do not cancel, because they do not point in opposite directions. As a result, the water molecule is polar.

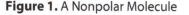

Figure 1. A Nonpolar Molecule

Figure 2. A Polar Molecule

We shall use valence shell electron-pair repulsion theory (VSEPR) as our guide in predicting molecular shapes. The basis of this theory is that pairs of electrons surrounding an atom try to get as far away from one another as possible to lessen electrostatic repulsion. In order to apply this theory, we must know how many pairs of electrons are around each atom in a molecule. It is therefore necessary to begin with the electron dot structure for the molecule.

Table 1 shows possible molecular shapes derived from VSEPR theory. Note how the electron pairs are placed as far apart as possible, whether the pair is a bonding pair or a nonbonding pair. Molecular *geometry* depends on the total number of substituents about the central atom. A *substituent* is any atom or any nonbonding pair of electrons. Notice that although the total number of electrons around the central atom is always eight (four pairs), the number of substituents can be less than four.

REPULSIVE DOTS

Remember that according to the VSEPR model, pairs of valence electrons strive to get as far apart as possible. With a multiple bond, the pairs cannot separate from one another, because they are all being shared by the same two nuclei. For the purposes of VSEPR, therefore, a multiple bond is treated as a single bond, and the noncentral atom taking part in the bond counts as *one* substituent.

Table 1

Electron Dot Structure	Number of Substituents	Number of Nonbonding Pairs	Molecular Geometry	Molecular Shape
B∴A∴B (with B top and bottom)	4	0	109° Tetrahedral	Tetrahedral
B∴A∴B (with B bottom)	4	1	109° Tetrahedral	Triangular pyramidal
B∴A∴B	4	2	109° Tetrahedral	Bent
B∴A∷B	3	0	120° Triangular planar	Triangular planar
A∷B	3	1	120° Triangular planar	Bent
B∷A∷B or B∴A∷B	2	0	180° Linear	Linear

Molecular *shape* depends on the placement of *atom substituents only*; nonbonding pairs are *not* considered when you are determining molecular shape. When there are no nonbonding pairs of electrons about a central atom, the molecular shape is identical to the molecular geometry. It is only when nonbonding pairs are present that a molecule's shape differs from its geometry.

Procedure

PART A: FROM ELECTRON DOT STRUCTURE TO MODEL

Using the molecular model kit provided by your instructor, build a model of each molecule shown in Table 2 of the report sheet. Then, sketch your molecules in the appropriate spaces in Table 2, using the bond notation used in column 5 of Table 1. Then, complete columns 3 and 4 of Table 2.

PART B: FROM MODEL TO ELECTRON DOT STRUCTURE

Use the three-dimensional models provided by your instructor to complete Table 3 of the report sheet.

Molecular Shapes Report Sheet

Table 2

Electron Dot Structure	Molecular Shape (Sketch of Three-Dimensional Model)	Name of Shape	Molecule Polar or Nonpolar?
H : N : H (with H below)			
H (top) H : C : Cl (H bottom)			
H : S : H			
O :: C :: O			
O :: S : O			
H : C ⋮⋮ N :			
H (top) C :: O (H bottom)			

REPULSIVE DOTS

Table 3

Model Number	Molecular Shape (Sketch of Model)	Electron Dot Structure	Name of Shape	Molecule Polar or Nonpolar?
1				
2				
3				
4				
5				
6				

Summing Up

1. Is BCl_3 a polar substance? Is NCl_3? Explain your answers.

_____

_____

_____

2. Is SCl_2 a bent molecule? Is H_2S? Explain your answers.

_____

_____

_____

CONCEPTUAL PHYSICAL SCIENCE	Activity

Molecular Models

Molecules by Acme

Purpose
Given a chemical formula, use the rules for how atoms bond to build a molecular model

Apparatus
molecular modeling kits *(Students should be able to work in groups of two or three.)*

Discussion
The three-dimensional shapes of molecules can be envisioned with the use of molecular models. There are certain things you must know, however, before you can assemble an accurate model of a molecule. First, you need to know what types of atoms make up the molecule and their relative numbers. This information is given by the molecule's chemical formula. The chemical formula for water, H_2O, for example, tells us that each and every water molecule is made of two hydrogen atoms and one oxygen atom. The second thing you need to know is how the atoms of the molecule fit together. For a water molecule there are several possibilities. One hydrogen atom might be bonded to both the second hydrogen atom and the oxygen atom (Figure 1a), or all three atoms might be bonded in the shape of a three-member ring (Figure 1b). We find through chemistry, however, that there are specific ways in which different types of atoms bond. Hydrogen, for example, forms only one bond, while oxygen forms two bonds. Knowing this, we can build a more reasonable model of a water molecule where both hydrogen atoms are bonded once to a central oxygen atom (Figure 1c).

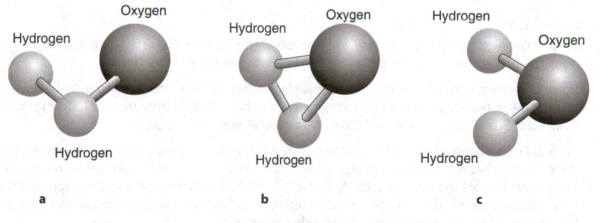

Figure 1. Which model of H_2O is correct?

The shape of a molecule is largely responsible for the physical and chemical properties of the molecule. If water were linear like carbon dioxide, for example, its boiling point would be close to that of carbon dioxide, -78°C, which would mean that Earth's oceans would be gaseous.

MOLECULES BY ACME

The Molecular Model

Molecular modeling kits represent different types of atoms with different-colored pieces. Hydrogen atoms, for example, are typically represented by white pieces, and oxygen atoms by red pieces. Also, the number of times an atom "prefers" to bond is indicated by the number of times the piece is able to attach to other pieces. Hydrogen atoms, for example, bond only once, so hydrogen pieces have only one site for attachment. Similarly, oxygen pieces have two sites for attachment. To build a molecule, the pieces representing atoms are connected by sticks (or springs). You will know when you have built a correct structure for a molecule when each atom is bonded the appropriate number of times (Table 1). In some instances, you may find it necessary to form multiple bonds between atoms. To see how this is done, consult the instructions to your modeling kit or your instructor. This is a hands-on play-as-you-investigate laboratory. Enjoy!

Table 1

Type of Atom	Atomic Symbol	Typical Color of Sphere	Number of Bonds* (holes)
Hydrogen	H	White	1
Carbon	C	Black	4
Nitrogen	N	Blue	3
Oxygen	O	Red	2
Chlorine	Cl	Green	1

* The number of times an atom tends to bond is related to its position in the periodic table. Consider, for example, the relative positions of carbon, nitrogen, oxygen, and chlorine. That these atoms are in adjacent columns and prefer 4, 3, 2, and 1 bonds, respectively, is not a coincidence. The periodic table is more than a table of facts. With further study you will find that the periodic table is highly organized—and a lot like a roadmap, because much information about an element can be told merely from its position within the periodic table.

Procedure

Step 1: Get a set of models. You may wish to work with a partner. The sets may not contain equal numbers of pieces, so you may occasionally need to borrow from other groups.

Step 2: Determine which colors should be used to represent the following elements: hydrogen, carbon, nitrogen, oxygen, chlorine, and iron. The number of bonds they are able to form should be as listed in Table 1. Enter the actual colors of the pieces you use in Table 2.

Step 3: Build models of each of the following molecules. Avoid forming triangular rings made of three atoms. They are strained and less stable. Check your structure with your teacher—in several cases there is more than one possibility. Complete Table 3 by drawing an accurate representation of your structure (follow the example of the water molecule). Before taking your models apart, answer the end-of-activity questions using your models as a guide.

1. Hydrogen gas: H_2	2. Oxygen gas: O_2	3. Nitrogen gas: N_2
4. Water: H_2O	5. Hydrogen Peroxide: H_2O_2	6. Ammonia: NH_3
7. Methane: CH_4	8. Dichloromethane: CH_2Cl_2	9. Chloroethanol: C_2H_5ClO
10. Carbon Dioxide: CO_2	11. Acetylene: C_2H_2	12. Ethanol: C_2H_6O
13. Acetic Acid: $C_2H_4O2_2$	14. Benzene: C_6H_6	15. Iron (III) Oxide: Fe_2O_3

Table 2

Element:	Hydrogen [H]	Carbon [C]	Nitrogen [N]	Oxygen [O]	Chlorine [Cl]	Iron (III)* [Fe]
Color:						

* For Iron, choose a piece that is able to bond in six directions at right angles of 90 degrees. The iron will form only three bonds such that three potential bonding sites remain unconnected.

Table 3

Hydrogen H_2	Oxygen O_2	Nitrogen N_2	Water H_2O	Hydrogen Peroxide H_2O_2

Ammonia NH_3	Methane CH_4	Dichloromethane CH_2Cl_2	Chloroethanol C_2H_5ClO	Carbon Dioxide CO_2

Acetylene C_2H_2	Ethanol C_2H_6O	Acetic Acid $C_2H_4O_2$	Benzene C_6H_6	Iron (III) Oxide Fe_2O_3

Summing Up

1. a. Atoms combine to form molecules in specific ratios. In a water molecule, for example, there are two hydrogens for every one oxygen. If this ratio were different—say two hydrogens to two oxygens—would the shape of the molecule also be different?

    _____

    b. Would you still have a water molecule?

    _____

2. a. How are the structures for methane and dichloromethane similar?

    _____

    _____

    b. How are they different?

    _____

    _____

3. How many *different* structures (configurations) are possible for the formula C_2H_5ClO? (*Hint*: There are more than two.)

    _____

4. Of the 15 models you made, which are linear?

    _____

5. Which molecules have multiple bonds between atoms?

    _____

6. Which of your structures is flat like a pancake?

    _____

CONCEPTUAL PHYSICAL SCIENCE **Activity**

How Atoms Bond and Molecules Attract   Paper Chromatography

Circular Rainbows

Purpose
To separate the different-colored components of black ink

Apparatus
a variety of black felt-tip pens
round filter paper or chromatography paper
a variety of solvents:
   water
   methanol (wood alcohol)
   ethanol (grain alcohol)
   isopropyl alcohol (rubbing alcohol)
   white distilled vinegar
   diethyl acetate (finger nail polish remover)
beakers or crucibles upon which to place the paper

Discussion
Black ink is made by combining many different-colored inks such as blue, red, and yellow. Together, these inks serve to absorb all the frequencies of light. With no light reflected, the ink appears black.

Procedure
It is easy to separate the components of an ink using a technique called *paper chromatography*. Place a concentrated dot of the ink at the center of a porous piece of paper, such as filter paper. Prop the marked paper on a small beaker or crucible, and then carefully add a drop of solvent such as water, white vinegar, or rubbing alcohol on top of the dot. Watch the ink spread outward with the solvent. Because the various components have differing attractions for the solvent, they travel with the solvent at different rates. Just before your drop of solvent is completely absorbed, add a second drop, then a third, and so on until the components have separated to your satisfaction. How the components separate depends on several factors, including your choice of solvent and your technique. Black felt-tip pens or pens used for overhead transparencies tend to work best, but you should experiment with a variety of different types of pens—even food coloring. It is also interesting to watch the leading edge of the moving ink under a microscope. Check for capillary action!

Summing Up

1. The different components of black ink not only have differing affinities for the solvent, they also have differing affinities for the paper, which is fairly polar. How would you expect a component that is ionically charged to behave while you were using a relatively nonpolar solvent, such as acetone? Would it readily travel with the solvent, or might it stay behind?

_____

2. Did you ever notice that not all "blue" inks are the same color blue? How might this be explained?

_____

CONCEPTUAL PHYSICAL SCIENCE	Experiment

Salt and Sand

Purpose
To plan a procedure for separating a mixture of salt and sand, modify the plan as needed during the procedure, and determine the percent of salt and sand in a mixture as accurately as you can

Apparatus
various salt-sand samples (about 10 grams each)
safety goggles
various pieces of laboratory equipment, depending on the procedure you develop

Discussion
Components of a mixture can be separated by their different physical properties. In this experiment you are to isolate salt and sand from a mixture. Knowing the mass of the isolated salt and the mass of the mixture, you can calculate the percent composition of the salt that was in the mixture. Likewise, knowing the mass of the isolated sand and the mass of the mixture, you can calculate the percent composition of the sand that was in the mixture. Your instructor will show you how to correctly use and care for laboratory equipment you may use, such as balances and hot plates. The procedure you follow for finding the percent compositions, however, is to be created by you. Before you do anything, therefore, sit down and write out what you think might be a good procedure to follow. For ideas, you might wish to discuss the possibilities with your classmates as well as your instructor. Label this procedure as your "Proposed Procedure."

Before you begin to work in the laboratory, show your written, step-by-step proposed procedure to your instructor for final approval. Don't be surprised if you have to modify your procedure as you move along. This is typical. You should also create, therefore, an "Actual Procedure" that shows what was actually done.

SAFETY!

Significant changes to your proposed procedure must be approved by your instructor. If you have approval to be working with a flame, remove all combustible materials, such as paper towels, from your work station. Eye protection is advised because it is not just what you're doing that poses a hazard. For example, others around you might accidentally shatter glassware, which could send glass pieces toward your face.

Procedure
Step 1: On a separate piece of paper or in a journal, write out a step-by-step procedure that you think will best allow you to find the amount of salt and sand within a mixture. Have your instructor approve of this procedure before moving on. Safety concerns and necessary equipment should be checked. In developing your procedure, keep in mind that you have to end up with the salt and the sand in separate containers so that you can measure the mass of each component.

Step 2: Obtain a sample of salt and sand from your instructor, and write down its identification number by your procedures and within question 1 of Summing Up.

Step 3: Begin following your proposed procedure, while recording your actual procedure. Write down numerical measurements, such as mass, neatly within one or more data tables. Include units with any number you write down.

Step 4: Use the following equations to calculate the mass percents of salt and sand in your mixture:

$$\text{Mass percent of salt} = \frac{\text{mass of salt}}{\text{mass of entire sample}} \times 100$$

$$\text{Mass percent of sand} = \frac{\text{mass of sand}}{\text{mass of entire sample}} \times 100$$

Because the mixture contains only two components, the total of the two percentages should be 100%. In reality, various errors may occur such that these two percentages **do not** add up to 100%.

Step 5: Obtain the true values for the mass percents from your instructor.

Summing Up

1. Enter your sample's identification number in this box.

2. What were the mass percents of salt and sand in your sample that you found through your procedure?

    Mass percent of salt: Mass percent of sand:

3. What were the true mass percents of salt and sand as reported to you by your instructor after you had completed your procedure?

    Mass percent of salt: Mass percent of sand:

4. What do you think was your most significant source of error in determining the mass percent of salt?

    _____

    _____

5. What do you think was your most significant source of error in determining the mass percent of sand?

    _____

    _____

CONCEPTUAL PHYSICAL SCIENCE	Activity

Mixtures   **Purification of Brown Sugar**

Pure Sweetness

Purpose
To isolate white sugar from brown sugar

Apparatus
brown sugar
kitchen cooking pot or 2000-mL beaker
empty glass bottle or 1000-mL beaker
kitchen knife or microspatula
small wide-mouthed jar (such as a baby food jar)
safety goggles
gloves for handling heated pot or beaker

Discussion
A supersaturated solution of brown sugar in water may be prepared by dissolving ample amounts of brown sugar in a small amount of boiling water. Upon cooling and after several days, a large number of crystals will have grown. These crystals will still be brown, but not quite as brown as the original brown sugar. Recrystallization of these brown crystals will result in crystals that are even less brown. Repeated recrystallization may result in crystals that appear white.

The formation of crystals is almost as much an art as it is a science. It is difficult, therefore, to guarantee that a particular procedure will always result in crystals of a given size and quality. With this in mind, expect to make some on-the-fly modifications to the following procedure.

This activity may last for several weeks, and you should keep a journal of the equipment and materials that you use and the actual procedure that you follow. Be sure to date every entry. Drawing sketches of your setup is also important.

Procedure
Step 1: In a cooking pot or large beaker, bring about 200 mL of water to a boil. Slowly add brown sugar to the boiling water with continuous stirring. Continue adding brown sugar until it becomes thick and just begins to froth. Do not overcook the brown sugar.

Step 2: Allow the solution to cool before pouring it into an empty glass bottle or into a 1000-mL beaker. Allow the syrup to stand for several days until a significant number of crystals have formed. The syrupy solution from which the crystals form is known as the *mother liquor*. Your crystals at this point will probably appear as tiny hard clumps.

Step 3: After there has been a significant amount of crystal growth, pour off the mother liquor. You may discard the mother liquor, but first note its close resemblance to molasses. In fact, that's exactly what it is! The brown color results from the many plant by-products it contains. The mother liquor will pour out very slowly. Consider leaving it inverted for several hours. The crystals should stay behind, stuck to the glass walls.

Step 4: Rinse the collected crystals with warm water to remove additional mother liquor. Rinse only briefly, because you want to avoid dissolving your crystals. After this quick rinse, isolate a few of these crystals and examine them closely.

Step 5: Use hot water to help collect all your crystals from your container. Use as little hot water as you can. Transfer your crystals with the small amount of water to a small pot or beaker, and apply heat until the crystals dissolve and some frothing is seen. You should end up with a syrupy solution once again.

Step 6: Allow the syrupy solution to cool before transferring it back to the cleaned glass bottle or beaker. A glass baby food jar also works well. Wait several days for more crystals to grow. This is called *recrystallization* because you are crystallizing a material that had already been crystallized. The effect is an increase in the crystal's purity.

Step 7: After crystal growth, the mother liquor of this solution can be poured out. The crystals that remain behind can be collected onto a towel, rinsed briefly with warm water, and quickly dried so that they retain crystal shape. If crystals still retain significant brown color, they may be recrystallized again using the general procedures given above.

Summing Up

1. How might you prove that commercial-grade white sugar still contains small amounts of molasses?

   _____

   _____

2. Which is more "pure": white sugar or brown sugar?

   _____

   _____

3. Which is more "natural": white sugar or brown sugar?

   _____

   _____

4. Write a statement regarding the quality vs. quantity of the sugar crystals you obtained from this activity.

   _____

   _____

CONCEPTUAL PHYSICAL SCIENCE	Experiment

Mixtures   Percent Sugar Determination

Sugar Soft

Purpose
To determine the sugar content of commercially sold soft drinks using a very simple hydrometer

Apparatus
plastic pipet (9 inches or comparable)
1/2-inch nut
sugar solutions (4%, 8%, 12%, 16%)
50-mL graduated cylinder
ruler (mm)
graph paper
various soft drinks (Try fruit juices and diet drinks as well!)

Discussion
A hydrometer is a flotation device used to measure the density of a liquid. The greater the density of the liquid, the higher the hydrometer floats. In this exploration, how high the hydrometer floats in four standard sugar solutions will be measured. The greater the sugar content of the solution, the greater its density, and hence, the higher the hydrometer floats. A *calibration curve* will be graphed showing the height of the hydrometer on the *y*-axis and the concentration of sugar on the *x*-axis. How high the hydrometer floats in various commercially prepared soft drinks, which are essentially sugar solutions with small amounts of other materials, will then be measured. Using the calibration curve, the sugar content of each soft drink may be estimated.

How high the hydrometer floats out of the water is the distance between its tip and the liquid surface (use units of millimeters). Make sure that the hydrometer is not held to the sides of the container—it should float as vertically as possible. The hydrometer must be rinsed and dried before each testing. Also, carbonated beverages must be "decarbonated" because bubbles that collect on the hydrometer affect its buoyancy. Your instructor may have already decarbonated some beverages by boiling them and then allowing them to cool.

Procedure

PART A: CONSTRUCTION OF A SIMPLE HYDROMETER
Step 1: Fill the pipet about half full with water and invert. All the water should run into the bulb.

Step 2: Slip the nut onto the stem end, and allow it to rest on the "shoulders" of the bulb as shown to the right.

Step 3: Test your hydrometer by placing it, bulb end down, in a 50-mL graduated cylinder containing 50 mL of water. The pipet should float with about 2.5 cm of the stem sticking out above the water. If it sticks out much more or less than this, either add water to the pipet or remove water from it.

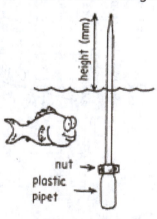

A simple hydrometer

SUGAR SOFT

Step 4: When you have adjusted the amount of water in the pipet bulb so that the stem sticks out about 2.5 cm, measure the exact height of the stem above the surface of the water in the graduated cylinder. Measure to the nearest millimeter, and record the height in Data Table 1.

Data Table 1

Concentration of Sugar Solution (%)	Height of Hydrometer above Liquid Surface (mm)
0 (plain water)	
4	
8	
12	
16	

Step 5: Remove the hydrometer from the graduated cylinder. *Being careful not to let any water spill out of the hydrometer or to let any water get in*, rinse the outside surface of the hydrometer well, and dry completely.

Step 6: Empty the graduated cylinder, rinse well, and dry completely.

Step 7: Add 50 mL of the 4% sugar solution to the graduated cylinder, place the hydrometer in the solution—bulb end down—and measure and record the height of the stem above the surface of the solution.

Step 8: Repeat Steps 5, 6, and 7 for 8%, 12%, and 16% sugar solutions, and enter all your data into Data Table 1.

Step 9: Plot the information in Data Table 1 into the large graph presented at the end of this exploration. Draw a straight line that best represents all the data points, as the example in Figure 1 shows. This straight line is also known as a calibration curve.

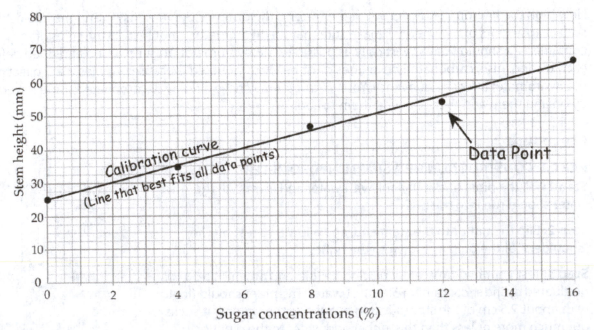

Figure 1. The calibration curve shown here is a line that best represents all of the data points. Note that the calibration curve does not necessarily touch all points.

PART B: DETERMINATION OF SUGAR CONTENT IN BEVERAGES

Step 1: Obtain soft drink and/or juice samples from your instructor.

Step 2: Following the procedure given in Part A, measure the stem height for each of your beverages, and record this information in Data Table 2.

Step 3: Use the calibration curve you prepared in Part A for your particular hydrometer to determine the sugar concentration in each of your beverages. To do this, find the point on the vertical axis that corresponds to the hydrometer stem height for your beverage. Figure 2 shows an example where the stem height is 52.5 mm. Draw a horizontal line that runs rightward from this point until it intersects the calibration curve. Then drop a vertical line from the intersection point down to the horizontal axis. The value at the point where this line intersects the horizontal axis—about 10.75% in the example shown here—is the sugar concentration of your beverage.

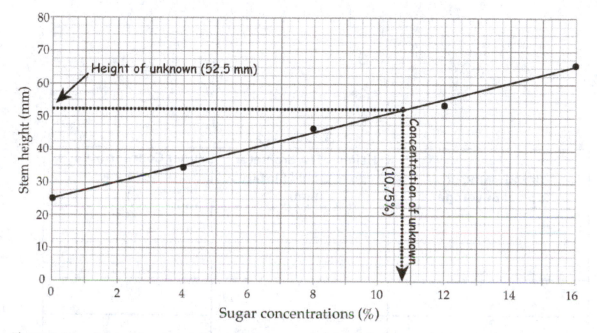

Figure 2. According to this sample calibration curve, a stem height of 52.5 mm translates into a sugar concentration of about 10.75%.

Step 4: Record the concentrations of all your beverages in Data Table 2.

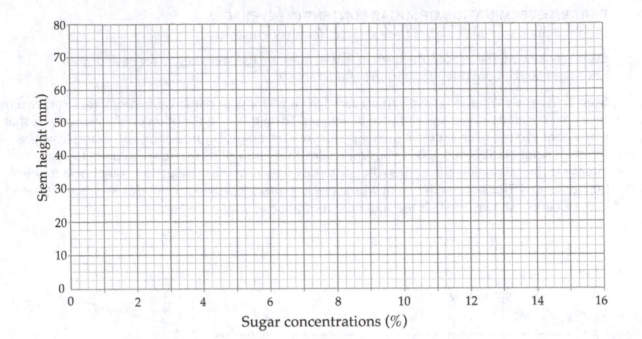

Data Table 2

Brand Name or Type of Beverage	Height of Hydrometer above Liquid Surface (mm)	Sugar Concentration of Beverage (%)

CONCEPTUAL PHYSICAL SCIENCE	Activity

How Chemicals React

Collection of a Gas

Bubble Round-up

Purpose
To isolate gaseous carbon dioxide by water displacement

Apparatus
baking soda   vinegar
two 250-mL Erlenmeyer flasks   1000-mL beaker
2 rubber stoppers (one with glass rod inserted through it)
tubing with a paper clip inserted in one end
ring stand with clamp   safety goggles

Discussion
A common method of collecting a gas produced from a chemical reaction is by the displacement of water. Bubbles of the gas are directed into an inverted flask filled with water. As the gas rises into the container, it displaces water. Once all the water is displaced, the flask may be sealed with a stopper. The physical and chemical properties of the gas can then be investigated. In this activity, baking soda (sodium bicarbonate, $NaHCO_3$), and vinegar (5% acetic acid, CH_3COOH), will react to form gaseous carbon dioxide, CO_2, which will be collected by water displacement. The other products of this reaction, water, H_2O, and sodium acetate, $CH_3COO^-Na^+$, form a liquid phase that remains in the reaction vessel.

$NaHCO_3$	+	CH_3CO_2H	—>	CO_2	+	H_2O	+	CH_3CO_2Na
sodium bicarbonate		acetic acid		carbon dioxide		water		sodium acetate

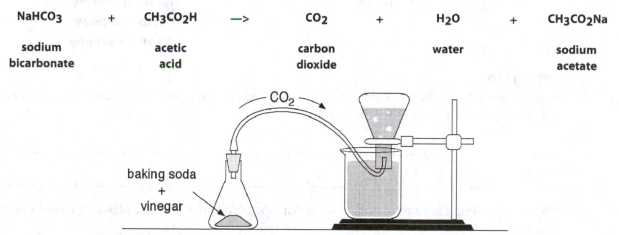

Figure 1. Baking soda (sodium bicarbonate) and vinegar (acetic acid) react to form gaseous carbon dioxide, CO_2.

Procedure
Step 1: Assemble the setup as shown in Figure 1 or in the manner presented to you by your teacher. A 250-mL Erlenmeyer flask is equipped with a rubber stopper through which a glass tube has been inserted. The glass tube is attached to a plastic tube that is forced into a J-shape on the opposite end by the bending of an inserted paper clip. A second 250-mL Erlenmeyer flask is filled with water and inverted into a 1000-mL beaker filled with 700 mL of water. The J-shaped end of the plastic tubing is then fixed below the lip of the inverted flask.

Step 2: Remove the stopper from the upright Erlenmeyer flask, and add about a tablespoon of baking soda (sodium bicarbonate) followed by a capful of vinegar (5% acetic acid). Allow the vigorous bubbling to settle down. Add a second capful of vinegar, and immediately stopper the upright flask. This will cause bubbles to fill the inverted flask. When bubbling stops, add another capful of vinegar to collect additional bubbles (carbon dioxide). Continue in this manner until the inverted beaker has been filled with carbon dioxide. *(Caution: Add vinegar only by the capful. Froth may overflow the flask if you add too much vinegar all at once.)*

Step 3: Stopper the inverted flask tightly to seal the carbon dioxide. Remove the flask from the clamp, and turn it right side up with the stopper still in place.

Step 4: Light a wooden match and, holding it, insert it into the flask containing the carbon dioxide. The flame should quickly extinguish. You might also light a small candle and slowly pour the carbon dioxide out of the flask onto the candle to extinguish the candle. This works because the carbon dioxide is heavier than air and will thus pour out of the flask.

Going Further

Step 5: Weigh the mass of an empty and dry 2-liter bottle with its cap. The mass of the air inside this bottle is about 2.3 grams. Subtract the mass of the air inside the bottle from the mass you measure to get the mass of only the bottle and its cap.

Step 6: Insert the J-hook end of the tube into the upright bottle, and pipe lots of carbon dioxide into the bottle. Make sure that the tube is dry and that no water drops get into the bottle. Because the CO_2 is heavier than air, it will settle to the bottom and displace the air, which will escape out the top. Continue filling the bottle with CO_2 until a match is extinguished when it is placed into the opening.

Step 7: Cap the bottle and measure its mass. Subtract the mass of the bottle and its cap from the mass you measure to get the mass of the carbon dioxide it contains. Calculate the density of the carbon dioxide by dividing its mass by its volume. Compare your experimental value to the actual density of carbon dioxide, which is 1.80 grams per liter.

Your calculated density
of carbon dioxide:

_____

**The actual density
of carbon dioxide:
1.80 grams per liter**

Summing Up

1. How certain are you that an empty 2-liter plastic soda bottle contains a volume of exactly two liters? How would you find out for sure?

_____

_____

2. Why was the flask containing baking soda not stoppered after the first capful of vinegar was added?

_____

_____

3. Why was the large beaker not initially filled to the brim?

_____

_____

CONCEPTUAL PHYSICAL SCIENCE	**Activity**

Two Classes of Chemical Reactions   **Red Cabbage Juice pH Indicator**

Sensing pH

Purpose
To isolate and use the pH-sensitive pigment of red cabbage

Apparatus
red cabbage
cooking pot
strainer
5 clear plastic cups or 100-mL beakers
safety goggles
water
household solutions: white vinegar, ammonia cleanser, grapefruit juice, club soda
household powders: baking soda, washing soda, borax, salt
9-volt battery

Discussion
The pH of a solution can be approximated by way of a pH indicator—any chemical whose color changes with pH. Many pH indicators can be found in plants. Red cabbage is one such plant.

Safety!

In this activity, you will be creating a broth of red cabbage in boiling water. Protect your eyes from the splattering of any liquid or the shattering of any glassware by wearing safety goggles. Protect your skin from scalding by handling the boiling water with care.

Procedure
Step 1: Shred about a quarter of a head of red cabbage. Boil the shredded cabbage in about 500 mL of water for about five minutes. Strain the cabbage while collecting the broth, which contains the pH indicator. Allow the broth to cool so that it can be safely handled.

Step 2: Fill five or more clear plastic cups or 100-mL glass beakers halfway with the broth. To each cup add a small amount of a household solution (about 10 mL) or powder (about a teaspoon). Examples of solutions you might use include white vinegar, ammonia cleanser, grapefruit juice, and club soda. Examples of powders you might use include baking soda, washing soda, borax, and salt.

Step 3: Note the changes in color, and use the following guidelines to estimate the pH of the solution.

Color of red cabbage indicator	dark red	pinkish red	light purple	light green	dark green
pH range	1–4	4–6.5	6.5–7.2	7.2–8	8–10

Enter in Table 1 the identity of the material you are testing and the estimated pH of the solution it makes.

Table 1

Household Solution or Powder	Estimated pH

Step 4: Add some fresh broth to a clear cup, and carefully submerge the terminals of a 9-volt battery into this broth. Hold on to the battery. Do not drop it into the broth. Look carefully for any change in color around either of the terminals. If you do not see color changes, you may have to dilute the solution. Look also for the formation of bubbles. Dry the 9-volt battery thoroughly after use, and return it to your teacher.

Summing Up

1. What color changes did you see upon submerging the terminals of the 9-volt battery in the red cabbage broth?

_____

2. Hydroxide ions, OH^-, are one of the products that form around the terminals of the 9-volt battery dipped into solution. At which terminal (the positive or the negative) did these hydroxide ions form?

_____

3. At which terminal on the 9-volt battery (the positive or the negative) did you see the formation of bubbles?

_____

CONCEPTUAL PHYSICAL SCIENCE	**Experiment**

Two Classes of Chemical Reactions

Titration

Upset Stomach

Purpose
To measure and compare the acid-neutralizing strengths of antacids

Apparatus

EQUIPMENT
buret stand with clamps
three 250-mL Erlenmeyer flasks
safety goggles
gloves
balance
mortar and pestle
weigh-dishes
plastic pipets
well-plates

CHEMICALS
0.50 M HCl solution
0.50 M NaOH solution
phenolphthalein indicator
various brands of antacid tablets (colorless)

Safety!
This is a fairly safe lab. Eye protection, however, must be worn at all times when you or your lab partners are working with solutions of hydrochloric acid, HCl, or sodium hydroxide, NaOH. You may also want to wear gloves to protect your skin, which may sting if it comes in contact with the hydrochloric acid or feel slippery if in contact with the sodium hydroxide. If either of these solutions gets onto your skin, tell your teacher and proceed to rinse your skin thoroughly with running water.

Discussion
Overindulging in food or drink can lead to acid indigestion, a discomforting ailment that results from the excess excretion of hydrochloric acid, HCl, by the stomach lining. An immediate remedy is an over-the-counter antacid, which consists of a base that can neutralize stomach acid. In this experiment, you will add an antacid to a simulated upset stomach. Not all the acid will be neutralized, however, so you will then determine the effectiveness of the antacid by determining the amount of acid that remains.

This is done by completing the neutralization with another base, sodium hydroxide, NaOH. The reaction between hydrochloric acid and sodium hydroxide is

HCl	+	NaOH	→	NaCl	+	H₂O
hydrochloric acid		sodium hydroxide		salt		water

The concentrations of the HCl and NaOH used in this experiment are the same. This means that the volume of NaOH needed to neutralize the remaining HCl in the "relieved stomach" is equal to the volume of HCl that is *not* neutralized by the antacid.

Procedure

PART A: PREPARING THE "UPSET STOMACH"

Step 1: Obtain a 250-mL Erlenmeyer flask.

Step 2: Have your teacher or another authorized individual deliver about 30.00 mL of a 0.50 M HCl solution to your 250-mL Erlenmeyer flask.

Step 3: Record in your data sheet the volume of 0.50 M HCl that your flask actually contains. For example, depending on the instrument used to deliver the HCl solution, your volume might be 29.93 mL or 29.9 mL or simply 30 mL.

Step 4: Add 2 drops of phenolphthalein pH indicator to the flask. No color change should be observed.

PART B: ADDING THE ANTACID

Step 1: Record the brand and active ingredient of your antacid in the data table.

Step 2: Crush and grind the antacid tablet with a mortar and pestle. (*Note*: Alka-Seltzer need not be crushed.)

Step 3: Carefully transfer all of the resulting powder to a weigh-dish, and then determine and record the mass of the weigh-dish plus powder.

Step 4: Carefully transfer the antacid from the weigh-dish to the "upset stomach" flask prepared in Part A, swirling for a few minutes while being careful not to spill. This flask now represents your "relieved upset stomach." (The solution should remain colorless.)

Step 5: Determine and record the mass of the empty weigh-dish.

Step 6: Calculate the mass of the antacid added to the flask representing the upset stomach.

PART C: COMPLETING THE NEUTRALIZATION

Step 1: Have your teacher show you the proper technique for delivering drops of liquid from a plastic pipet. This involves holding the pipet firmly while gently squeezing the bulb so that you can control drops coming out of the pipet one at a time. Practice your technique by delivering drops of water over a sink.

Step 2: Use a clean pipet to transfer 20 drops of the solution in the "relieved upset stomach" flask to each of five wells in the well-plate.

Step 3: Fill a clean pipet with 0.50 M NaOH solution, and then carefully and slowly add the NaOH dropwise to one of the wells. Keep a careful count of the number of drops added. You will see the "stomach" solution in the well turn pink right where the NaOH drops into it. This pink color will disappear as soon as the NaOH mixes in with the "stomach" solution. As you continue to add drops, however, the pink color will linger for longer periods of time. Keep adding drops until the solution retains a light pink color for at least 30 seconds. Do NOT count any drops you add beyond this constant light pink color. Once the pink color remains, the total amount of acid within the "upset stomach" solution has been neutralized.

Step 4: In the data table, record the minimum number of drops of NaOH required to neutralize the "stomach acid" solution. This is equal to the number of drops required to obtain the constant light pink color. Repeat Step 3 for the remaining wells.

Step 5: Repeat the entire procedure, starting from Part A, for two other brands of antacid.

Data

PARTS A AND B: PREPARING THE "UPSET STOMACH" AND ADDING THE ANTACID

	Antacid 1	Antacid 2	Antacid 3
1. Antacid used	_____	_____	_____
2. Active ingredient	_____	_____	_____
3. Volume of HCl added to Erlenmeyer flask	_____ mL	_____ mL	_____ mL
4. Number of drops of phenolphthalein added	_____drops	_____drops	_____drops
5. Mass of weigh-dish with crushed antacid	_____ g	_____ g	_____ g
6. Mass of weigh-dish after antacid was transferred	_____ g	_____ g	_____ g
7. Mass of antacid added to "stomach"	_____ g	_____ g	_____ g

PART C: COMPLETING THE NEUTRALIZATION

Antacid 1	Well 1	Well 2	Well 3	Well 4	Well 5		Total
1. Drops of "relieved stomach" fluid	20	20	20	20	20	=	100
2. Drops of NaOH added to complete neutralization						=	
3. Drops of "stomach acid" neutralized by NaOH						=	
4. Drops of "stomach acid" neutralized by antacid						=	

Antacid 1	Well 1	Well 2	Well 3	Well 4	Well 5		Total
1. Drops of "relieved stomach" fluid	20	20	20	20	20	=	100
2. Drops of NaOH added to complete neutralization						=	
3. Drops of "stomach acid" neutralized by NaOH						=	
4. Drops of "stomach acid" neutralized by antacid						=	

Antacid 1	Well 1	Well 2	Well 3	Well 4	Well 5	Total
1. Drops of "relieved stomach" fluid	20	20	20	20	20	100
2. Drops of NaOH added to complete neutralization						
3. Drops of "stomach acid" neutralized by NaOH						
4. Drops of "stomach acid" neutralized by antacid						

Summing Up

1. List the antacids you tested in order of neutralizing strength, strongest first:

   strongest _____ > _____ > _____ weakest

2. List the antacids you tested in order of the masses of the tablets, most massive first:

   most massive _____ > _____ > _____ least massive

3. Divide the total number of drops of "stomach acid" neutralized by each antacid by the mass of the antacid that was added to the "stomach":

   Antacid 1: _____ drops/gram

   Antacid 2: _____ drops/gram

   Antacid 3: _____ drops/gram

4. List the antacids in order of neutralizing strength based on the number of drops of acid neutralized for every gram of antacid:

   strongest _____ > _____ > _____ weakest

5. What would be the effect on the neutralizing strength for an antacid if the following errors were made?

   a. A student spills some of the crushed antacid as it is transferred to the "stomach" flask:

   _____

   b. A student "overshoots" the number of drops of sodium hydroxide so that the solution turns a dark pink rather than a light pink. The student then counts this number of extra drops in the Data Table:

   _____

CONCEPTUAL PHYSICAL SCIENCE	**Activity**

Two Classes of Chemical Reactions   **Battery Basics and a Battery Puzzle**

The Lemon Electric

Purpose
To use the voltmeter function of a digital multimeter to find the relationship between the voltage of a battery and its size and to create a battery using simple materials

Apparatus
digital multimeter (DMM)   C- or D-cell alkaline battery
ignition dry cell (No. 6) battery (if available)   small alkaline battery (N- or AAAA-cell)
lemon half or wedge (or equivalent)   2 galvanized nails
plastic plate   access to water and paper towel (for clean-up)

Going Further
pennies   aluminum foil

Discussion
Now that so many devices are electronic, more sources of energy are required to power them. The most common of these is the battery, composed of a combination of cells. In this activity, you'll consider the common battery and a bit of its history.

Procedure
PART A: SIZE AND VOLTAGE
Step 1: Arrange the multimeter to measure volts (DC). Set the dial to "V" with a range of perhaps 2 V, 20 V, or 200 V. Connect the black multimeter lead to the "common" (often black) jack. Connect the red lead to the red jack labeled with a V.

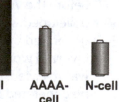

Digital   C- or   Ignition Cell   AAAA-   N-cell
Multimeter   D-cell   cell

Step 2: a.   Measure the voltage of the C- or D-cell.

   V = _____

b.   The large No. 6 ignition battery has a volume more than 20 times that of the C- or D-cell battery. Measure and record the voltage of the large battery.

   V = _____

c.   An N-cell battery has about 1/8 the volume of the C- or D-cell. An AAAA-cell has about 1/20 the volume of a C- or D-cell battery. Measure and record the voltage of the small battery.

   V = _____

Step 3: Consider your observations in the previous step. What relationship—if any—is apparent between the size of a battery and its voltage?

_____

_____

PART B: BACKGROUND
Read the following to learn the basic nature of the battery.

SCIENCE GETS THE BOOT FROM ITALY
When Galileo was sentenced to house arrest for his "crimes" against the Roman Catholic Church, the Scientific Revolution fled Italy for the more hospitable regions of France and England. Later, in England, Isaac Newton advanced the understanding and mathematical description of the dynamics of motion, gravity, and optics. In British America, Benjamin Franklin advanced the understanding of static electricity, and in France, Charles Coulomb applied Newtonian mathematical rigor to the description of electrostatic forces.

LUIGI AND ALESSANDRO: SCIENCE RETURNS TO ITALY
A century and a half after Galileo's death, something of scientific importance was to develop in Italy. During the 1780s, biologist Luigi Galvani performed experiments at the University of Bologna involving electrical impulses and frogs. It had been found that a charge applied to the spinal cord of a frog could generate muscle spasms throughout its body. Charges could make frog legs jump, even if the legs were no longer attached to a frog. While cutting a frog leg, Galvani's steel scalpel touched a brass hook that was holding the leg in place. The leg twitched. Further experiments confirmed this effect. Galvani was convinced that he was seeing the effects of what he called "animal electricity," the life force within the muscles of the frog.

At the University of Pavia, Galvani's colleague Alessandro Volta was able to reproduce the results, but was skeptical of Galvani's explanation. Volta, a former high school physics teacher, found that it was the presence of two dissimilar metals—not the frog leg—that was critical. In 1800, after extensive experimentation, he developed the voltaic pile. The original voltaic pile consisted of a stack of zinc and silver discs and between alternate discs, a piece of cardboard that had been soaked in saltwater. A wire connecting the bottom zinc disc to the top silver disc could produce repeated sparks. No frogs were harmed in the production of a voltaic pile.

Before the voltaic pile was developed, sparks had to be generated by friction. That involved work. When the charge was released, another spark could be generated only by more frictional work. The voltaic pile provided a continuous source of charge flow. No work had to be done! Volta's pile is widely regarded as the first battery.*

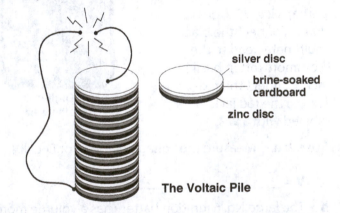

silver disc
brine-soaked cardboard
zinc disc

The Voltaic Pile

The terms **galvanic, galvanize,** and **galvanometer** (among others) honor Luigi Galvani. The unit of electric potential, the **volt,** was named in honor of Alessandro Volta. Electric potential, itself, is commonly referred to as **voltage.**

*An item found in an archeological excavation beneath the city of Baghdad consisted of a ceramic vessel enclosing a copper cup, which, in turn, enclosed a smaller silver cup. This reveals that a battery may have existed in ancient Persia. Unfortunately, it is not known why it was constructed or what it was used for. However, these details are known in the case of the voltaic pile.

What were the **critical components** of Galvani's and Volta's batteries?

_____

_____

PART C: THE LEMON BATTERY

Step 1: Use the lemon, nails, **_and your resourcefulness and creativity_** to design a battery that will register 0.5 V or more. (Don't worry if the reading is negative; what's important is that the magnitude is 0.5 or greater.) When you think you have the correct design, ask the instructor to let you make a voltage measurement using the digital multimeter.

When using the multimeter, be sure to press the leads firmly against the elements of your battery. A gentle touch won't ensure the complete connection you'll need to register an accurate reading.

Always keep in mind the critical elements of the battery!

Step 2: When you achieve success,

a.   record your qualifying voltage:_____.

b.   describe your successful design. Where did your resourcefulness and creativity come into play?

_____

_____

_____

Step 3: Draw two diagrams in the space below. Draw one for a configuration that doesn't work and one for a configuration that does work. Title each diagram appropriately.

Going Further

1. Given access to other materials (such as aluminum foil, pennies, and paper towel) in addition to your lemon, can you create a voltaic pile?

People with dental fillings can make a battery in their mouth if they accidentally bite aluminum foil. It's painful, so they don't tend to do it on purpose!

   a. How do you properly create a voltaic pile?

_____

_____

_____

   b. What's the highest voltage you can attain?

_____

CONCEPTUAL PHYSICAL SCIENCE	Activity
Two Classes of Chemical Reactions	**Percent Oxygen in Air**

Tubular Rust

Purpose
To determine the percent oxygen in air

Apparatus

steel wool
pencil
centimeter stick and rubber band
ring stand and test tube clamp

test tube (15 cm)
250-mL beaker
white distilled vinegar (5% acetic acid)

Discussion

This activity takes advantage of the rusting of iron by oxygen to determine the percent oxygen in the air. Iron is placed in an air-filled test tube, which is then inverted in water. As the iron reacts with the oxygen, the pressure inside the test tube decreases, and atmospheric pressure pushes water up into the tube. The decrease in the volume of air in the test tube is a measure of the depletion of its oxygen content. By measuring the quantity of air in the test tube before and after the rusting of the iron, it is possible to calculate the percent oxygen in air by volume.

Procedure

Step 1: Measure about one gram of steel wool that has been pulled apart to increase its surface area. To promote rusting, dip the wool into some white distilled vinegar. Shake off the excess vinegar, and push the steel wool halfway into a test tube 15 cm in length using a pencil. The wool should be packed as loosely as possible to maximize its surface area, but packed tight enough so that it does not fall out when the tube is inverted.

Step 2: Attach a lightweight centimeter stick to the test tube using a rubber band. Position the stick so that the 0-mm mark is toward the open end. Carefully invert the test tube into the beaker, which should be filled halfway with water. Clamp the test tube to a ring stand, and adjust its height so that the lip is no more than a centimeter under the water. Adjust the ruler so that the 0-mm mark is even with the water level *inside* the tube.

Step 3: Water will climb up into the test tube as the rusting proceeds. As this occurs, lower the test tube deeper into the beaker so that the water levels outside and inside the test tube remain even. Begin to read and record the water level inside the test tube at three-minute intervals until it stops rising.

Step 4: Plot a graph of the water level inside the test tube vs. time.

Step 5: Calculate the fraction of oxygen in air as the maximum height of the water inside the tube divided by the total length of the tube. Multiply by 100 to obtain a percentage.

Step 6: Leave the test tube submerged until your next class period to see whether the water level inside the tube goes up even higher. If it does, recalculate the fraction of oxygen in air.

Summing Up

1. What is your experimental percent oxygen in the air?

_____

_____

2. Why is it important that the water levels inside and outside the test tube remain even?

_____

_____

3. Would it take a longer time or a shorter time for the oxygen to be depleted if the steel wool were packed tightly at the bottom of the test tube? Briefly explain.

_____

_____

CONCEPTUAL PHYSICAL SCIENCE	Activity

Organic Compounds   **Preparation of Fragrance Esters**

Smells Great!

Purpose
To prepare the wintergreen fragrance

Apparatus
test tube (greater than 10 cm in length)
concentrated sulfuric acid in dropper bottle
safety goggles
methanol
salicylic acid

Discussion
Many flavorings and fragrances, both natural and artificial, are a class of organic molecules called *esters*. These molecules contain a carbonyl group bonded to an oxygen atom that is bonded to a carbon atom (see the organic chemistry chapter of your text). Esters can be prepared in the laboratory by reacting alcohols with carboxylic acids (Figure 1).

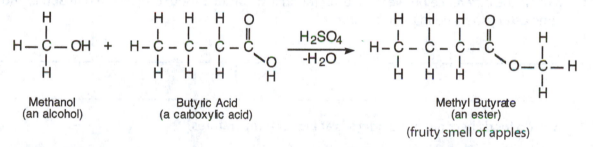

Methanol   Butyric Acid   Methyl Butyrate
(an alcohol)   (a carboxylic acid)   (an ester)
                                                              (fruity smell of apples)

Figure 1. An example of the formation of an ester from an alcohol and a carboxylic acid.

In this activity, you and your class will prepare small but fragrant quantities of methyl salicylate, which is the ester formed by reacting methanol with salicylic acid (Figure 2). This ester gives us the smell of wintergreen.

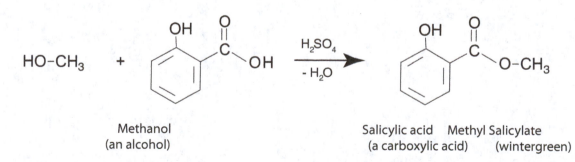

Methanol   Salicylic acid   Methyl Salicylate
(an alcohol)   (a carboxylic acid)   (wintergreen)

Figure 2. Methanol and salicylic acid react to form methyl salicylate.

> **Safety!**
> This lab requires the use of concentrated sulfuric acid, H_2SO_4. Only one drop per test tube will be used, but because this acid is concentrated, these drops can be very harmful to skin. Your instructor will be dispensing the drop of concentrated sulfuric acid into your test tube. Handle your samples with care. They may smell good, but do NOT taste them because they may still contain residual amounts of sulfuric acid. Wear safety goggles at all times.

Procedure

Step 1: Place 0.2 g of salicylic acid into a clean, dry 10-cm test tube.

Step 2: Add 15 drops of methanol, and lightly tap the bottom of the test tube to mix the contents.

Step 3: Allow your teacher to add a drop of concentrated sulfuric acid directly into the mixture within your test tube. Again, lightly tap the bottom of the tube to mix the contents.

Step 4: Place your test tube into a beaker of boiling water for three minutes. Look for white fumes coming from the mouth of the test tube.

Step 5: Remove the test tube from the boiling water, and allow it to cool enough so it can be held.

Step 6: Gently wave your hand over the top of the test tube to waft air from the test tube to your nose. Do this until you detect the smell of wintergreen.

Step 7: When you are finished, follow the clean-up instructions given to you by your teacher.

Summing Up

1. What is the difference between an ester found in a natural product, such as a pineapple, and the same ester produced in the laboratory?

   _____

   _____

2. Why do foods become more odorous at higher temperatures?

   _____

   _____

CONCEPTUAL PHYSICAL SCIENCE | Activity

Organic Compounds   **Densities of Organic Polymers**

Name That Recyclable

Purpose
To identify a variety of unknown recyclable plastics on the basis of their densities

Apparatus
recyclable plastics (PET, HDPE, LDPE, PP, PS)
solutions:
   95% ethanol and water (1:1 by volume)
   95% ethanol and water (10:7 by volume)
   10% NaCl in water

Discussion
There are many different types of plastics and many of them are recyclable. Ultimately, plastics need to be sorted according to their composition. For this reason, plastics are coded with a number within the recycling arrow logo. The initials of the type of plastic may also appear. For example, the plastic used to make 2-liter soft-drink bottles is polyethylene terephthalate (PET). This plastic has the following recycling code:

Plastics can also be identified based on their densities, which is useful if the recycling code is unreadable or absent. In this activity, you are to identify pieces of unknown plastics on the basis of their densities.

Procedure
Step 1: Use the information in Table 1 to develop a separation scheme by which you will be able to identify all your unknown pieces of plastic.

Step 2: Use the solutions provided to identify each unknown piece of plastic:

Unknown No.:					
Identity:					

Table 1

Recyclable Plastic	Density Compared to...			
	Water	Ethanol and Water (1:1)	Ethanol and Water (10:7)	10% NaCl in Water
△ 1 PET (or PETE) Polyethylene terephthalate	Greater	Greater	Greater	Greater
△ 2 HDPE High-density polyethylene	Less	Greater	Greater	Less
△ 4 LDPE Low-density polyethylene	Less	Less	Greater	Less
△ 5 PP Polypropylene	Less	Less	Less	Less
△ 6 PS Polystyrene	Greater	Greater	Greater	Less

Summing Up

1. What other physical properties might be used to identify unknown pieces of plastic?

_____

_____

2. You have just been given a thousand pounds of recyclable polystyrene and polypropylene plastic pieces all mixed together. Suggest how you might quickly separate the different types of plastic from one another.

_____

_____

CONCEPTUAL PHYSICAL SCIENCE	**Activity**

Rocks and Minerals   **Crystals**

Crystal Growth

Purpose
To observe the growth of crystals from a melt and from a solution

Apparatus
thymol
sodium chloride
sodium nitrate
potassium aluminum sulfate (alum)
copper acetate
glass slide or clear glass plate
forceps
petri dish
hot plate
microscope

Discussion
A **mineral** is a naturally formed inorganic solid composed of an ordered array of atoms. The atoms in different minerals are arranged in their own characteristic ways. This systematic arrangement of atoms is known as a mineral's crystalline structure, which exists throughout the entire mineral specimen. If crystallization occurs under ideal conditions, it will be expressed in perfect crystal faces.

I. GROWTH OF CRYSTALS FROM A MELT
Most minerals originate from a magma melt. Crystalline minerals form when molten rock cools. The temperature at which minerals crystallize is very high (well above the boiling point of water), making direct observation difficult. In this experiment we will use thymol, an organic chemical that crystallizes near room temperature. Although thymol is different from a magma, the crystallization principles exhibited are similar.

Safety
Thymol is not poisonous but may cause skin and eye irritation. Be safe: use forceps to handle thymol.

Procedure

PART A: SLOW COOLING WITHOUT "SEED" CRYSTALS
Set the hot plate to low heat. Place a petri dish containing a small amount of crystalline thymol on the hot plate until all the crystals are melted. Allow the melt to heat for one to two minutes, and then set it aside to cool slowly. Do not disturb it during cooling. This melt will be examined at the end of the lab session.

CRYSTAL GROWTH

PART B: SLOW COOLING WITH "SEED" CRYSTALS

Repeat the procedure in Part A, but transfer the petri dish to the stage of a microscope as soon as the thymol is melted. Add several (four or five) "seed" crystals to the melt, and observe it under the microscope. As the melt cools, you will see the crystals begin to grow.

Describe the manner of crystal growth. Think about the rate of growth, the direction of growth, the crystal faces, and the effect of limited space on crystal shape. What role do the "seed" crystals play in initiating crystal growth?

_____

_____

Make a sketch of the final crystalline solid. The crystals produced in this experiment are similar to the crystals found in common igneous rocks. Both are interlocking crystals. Compare the thymol crystals to those in a rock specimen of granite. Sketch the texture of the granite. How do the thymol crystals compare with those in granite?

_____

_____

PART C: RAPID COOLING

Repeat the procedure in Part A, but transfer the petri dish to the top of an ice cube and let it cool for 30 seconds. The melt will cool very rapidly. Quickly move the petri dish to the microscope, and observe the nature of the crystal growth. Time permitting, repeat this procedure until you are sure of your observations and can state your general conclusions.

_____

_____

_____

PART D: EXAMINATION OF CRYSTALS FROM PROCEDURE IN PART A

Examine the thymol crystals from the first part of this exercise. Observe the crystal size. What effect does the rate of cooling have on the crystal size?

_____

_____

II. GROWTH OF CRYSTALS FROM SOLUTION

Many minerals are precipitated from aqueous solutions by evaporation. Crystal growth of such minerals can be observed in the laboratory by evaporating prepared concentrated solutions.

Procedure

From your instructor, obtain concentrated solutions of sodium chloride, sodium nitrate, potassium aluminum sulfate (alum), and copper acetate.

Place a drop of each solution on a microscope slide. Label each slide accordingly. As the water evaporates from the slide, crystals of each compound will appear. With time and continued evaporation, the crystals will grow larger.

Summing Up

1. How does crystallization from an aqueous solution compare to crystallization from a melt?

_____

_____

_____

2. Do the crystals precipitated from a solution have unique crystal forms?

_____

_____

_____

3. According to your observations in this activity, do you think that crystal forms provide a good reference for mineral identification?

_____

_____

_____

4. According to your observations in this activity, what is the relationship between cooling and crystal size?

_____

_____

_____

CONCEPTUAL PHYSICAL SCIENCE	Activity

Rocks and Minerals **Classification of Rock-forming Minerals**

What's That Mineral?

Purpose
To observe the physical properties of various mineral samples, and identify them by a systematic procedure

Apparatus
mineral collection
streak plate (non-glazed porcelain plate)

hardness set—glass, steel knife, copper penny,
dilute hydrochloric acid (HCl)

Discussion
A mineral is a naturally formed inorganic solid with a characteristic chemical composition and crystalline structure. The different combinations of its elements and arrangements of atoms determine the physical properties of a mineral: its shape, the way it reflects light, its color, its hardness, and its mass.

I. The physical properties dependent on a mineral's chemical composition include luster, streak, color, specific gravity, and reaction with acid.

The **luster** of a mineral is the appearance of its surface when it reflects light. Luster is independent of color; minerals of the same color may have different lusters, and minerals of the same luster may have different colors. Mineral luster is classified as either metallic or nonmetallic.

TEST FOR LUSTER	
Metallic minerals	**Nonmetallic minerals**
a. are usually gold, silver, or black	a. are usually not gold or silver
b. are usually shiny, polished	b. are usually shiny to dull
c. are usually opaque	c. are usually transparent, translucent, or opaque
d. always have a streak	d. rarely have a streak

The **streak** of a mineral is the color of its powdered form. We can see a mineral's streak by rubbing it across a non-glazed porcelain plate. Minerals with a metallic luster generally leave a dark streak that may be different from the color of the mineral. For example, the mineral hematite is normally reddish brown to black but always streaks red. Magnetite is normally iron-black but streaks black. Limonite is normally yellowish brown to dark brown but always streaks yellowish brown. Minerals with a nonmetallic luster either leave a light streak or leave no streak at all.

TEST FOR STREAK
1. Scrape the mineral across a non-glazed porcelain plate.
2. Blow away excess powder.
3. The color of the powder is the streak.

Although **color** is an obvious feature of a mineral, it is not a reliable means of identification. When used with luster and streak, color can sometimes aid in the identification of metallic minerals. Nonmetallic minerals may occur in a variety of colors or be colorless. Therefore, color is not used for the identification of nonmetallic minerals. In this exercise, we will only differentiate between light-colored and dark-colored minerals.

Specific gravity (s.g.) is the ratio of the mass (or weight) of a substance to the mass (or weight) of an equal volume of water. Metallic minerals tend to have a higher specific gravity than nonmetallic minerals. For example, the metallic mineral gold (Au) has a specific gravity of 19.3, whereas quartz (SiO_2), a nonmetallic mineral, has a specific gravity of 2.65.

The **reaction to acid** is an important chemical property often used to identify carbonate minerals. Carbonate minerals effervesce (fizz) in dilute hydrochloric acid (HCl). Some carbonate minerals react more readily with HCl than do others. For instance, calcite ($CaCO_3$) strongly effervesces when exposed to HCl, but dolomite ($CaMg(CO_3)_2$) doesn't react unless it is scratched and powdered.

II. The physical properties dependent on a mineral's crystalline structure include hardness, cleavage and fracture, crystal form, striations, and magnetism.

The resistance of a mineral to being scratched (or its ability to scratch) is a measure of the mineral's **hardness**. The varying degrees of hardness are represented on the Mohs scale of hardness. For this activity, we are concerned with the hardness of some common objects.

Cleavage and fracture are useful guides for identifying minerals. **Cleavage** is the tendency of a mineral to break along planes of weakness. Planes of weakness depend on crystal structure and symmetry. Some minerals have distinct cleavage. Mica, for example, has perfect cleavage in one direction and breaks apart to form thin, flat sheets. Calcite has perfect cleavage in three directions and breaks to produce rhombohedral faces that intersect at 75° degrees. A break other than along cleavage planes is a **fracture**.

MOHS SCALE OF HARDNESS		
Mineral	**Scale Number**	**Common Test Object**
Diamond	10	
Corundum	9	
Topaz	8	
Quartz	7	Steel file
Feldspar	6	Window glass
Apatite	5	Pocket knife
Fluorite	4	
Calcite	3	Copper penny
Gypsum	2	Fingernail
Talc	1	

TEST FOR HARDNESS
1. Place a glass plate on a hard, flat surface.
2. Scrape the mineral across the glass plate.
3. If the glass is scratched, the mineral is harder than glass.
4. If the glass is not scratched, the mineral is softer than glass.
5. If the mineral is softer than glass, test to see whether it is harder than a copper penny. Scrape the mineral across the penny.
6. If the mineral is softer than the penny, test to see whether it is harder than your fingernail. Try to scratch the mineral with your fingernail.

Every mineral has its own characteristic **crystal form**. Some minerals have such a unique crystal form that identification is relatively easy. The mineral pyrite, for example, commonly forms as intergrown cubes, and quartz commonly forms as six-sided prisms that terminate in a point. Most minerals, however, do not exhibit their characteristic crystal form. Perfect crystals are rare in nature because minerals typically grow in cramped, confined spaces.

CLEAVAGE PATTERNS

Number of Cleavage Directions	Shape	Number of Flat Surfaces	Sample Illustration
1	Flat sheets	2	
2 at 90°	Rectangular cross section	4	
2 not at 90°	Parallelogram cross section	4	
3 at 90°	Cube	6	
3 not at 90°	Rhombohedron	6	
Fracture	Irregular shape	0	

Some minerals have grooves on their cleavage planes. These grooves, called **striations**, can be used to differentiate between feldspar minerals. Plagioclase feldspars have straight, parallel striations on one cleavage plane. Orthoclase feldspars have lines that resemble striations but are actually color variations within the mineral. These grooves are not straight, and they are not parallel to each other. Instead, these "striations" make a criss-cross pattern.

Some minerals exhibit magnetism. To test for magnetism, simply expose the mineral to a small magnet or a compass.

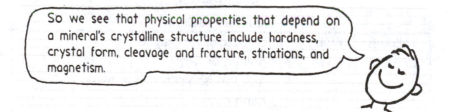

So we see that physical properties that depend on a mineral's crystalline structure include hardness, crystal form, cleavage and fracture, striations, and magnetism.

MINERALS WITH METALLIC LUSTER

Streak Color	Properties	Mineral
Black to gray	silver gray 3 directions of cleavage at 90° cubic crystals. hardness = 2.5 specific gravity = 7.6	Galena
Black to gray	black to dark gray magnetic hardness = 6 specific gravity = 5	Magnetite
Black to gray	gray to black marks paper feels greasy hardness = 1–2 specific gravity = 2.2	Graphite
	golden yellow may tarnish purple hardness = 4 specific gravity = 4.2	Chalcopyrite
Black to greenish black	brass yellow may tarnish green cubic crystals striations hardness = 6 specific gravity = 5.2	Pyrite (fool's gold)
Reddish brown to black	silver to gray may tarnish reddish brown hardness = 5–6 specific gravity = 5	Hematite
Yellowish brown to reddish brown	black to silver gray to golden brown may tarnish yellowish brown hardness = 5.5 specific gravity = 4	Limonite

Some are interesting...
some are more interesting!

MINERALS WITH NONMETALLIC LUSTER, DARK-COLORED

Hardness	Cleavage	Properties	Mineral
Harder than glass		black to blue gray 2 directions of cleavage, not quite at 90° striations on one cleavage plane hardness = 6 specific gravity = 2.7	Plagioclase
		dark green to black 2 directions of cleavage at nearly 90° no striations hardness = 6 specific gravity = 3.5	Pyroxene
		dark green to black 2 directions of cleavage intersecting at 60° and 120° no striations hexagonal crystals hardness 5.5 specific gravity = 3.6	Amphibole (hornblende)
Softer than glass	Present	brown to dark green to black 1 direction of cleavage producing thin sheets translucent hardness = 3 specific gravity	Biotite
		green to dark green 1 direction of cleavage producing thin, curved sheets greasy luster hardness = 2.5 specific gravity = 2.9	Chlorite

Earth science is down to earth!

MINERALS WITH NONMETALLIC LUSTER, DARK-COLORED

Hardness	Cleavage	Properties	Mineral
Harder than glass		olive green to black no streak glassy luster conchoidal fracture hardness = 7 specific gravity = 2.6	Olivine
		light to dark gray no streak glassy, transparent, translucent conchoidal fracture hexagonal crystals hardness = 7 specific gravity = 2.6	Quartz
		deep red to brown no streak translucent conchoidal fracture isometric crystals hardness = 7 specific gravity = 4	Garnet
Variable	Absent	red reddish brown streak opaque, earthy uneven fracture hardness = 1.5–5.5 specific gravity = 5.2	Hematite

When I dream of geology I have rocks in my head!

MINERALS WITH NONMETALLIC LUSTER, LIGHT-COLORED

Hardness	Cleavage	Properties	Mineral
Harder than glass		pale orange-pink, white	Orthoclase feldspar
		green, brown 2 cleavage directions at nearly 90° no striations color lines on cleavage planes hardness = 6 specific gravity = 2.6	
		white to blue-gray 2 directions of cleavage not quite at 90° striations on one cleavage plane hardness = 6 specific gravity = 2.7	Plagioclase feldspar
Softer than glass	Present	colorless to white 3 directions of cleavage at 90° soluble in water hardness = 2.5 specific gravity	Halite
		colorless to white 1 direction of cleavage hardness = 2 (easily scratched with fingernail) specific gravity = 2.3	Gypsum
		colorless to white or yellow 3 directions of cleavage not at 90° (rhomb-shaped) translucent to transparent strong reaction to acid hardness = 3 specific gravity = 2.7	Calcite

Hmmm...gold maybe?

WHAT'S THAT MINERAL?

MINERALS WITH NONMETALLIC LUSTER, LIGHT-COLORED

Hardness	Cleavage	Properties	Mineral
Softer than glass	Present	white, gray, pink 3 directions of cleavage not at 90° (rhomb-shaped) opaque reacts to acid when powdered hardness = 3.5 specific gravity = 2.9	Dolomite
		yellow, blue, green, violet 4 directions of cleavage transparent to translucent cubic crystals hardness = 4 specific gravity = 3.2	Fluorite
		colorless to pale green 1 direction of cleavage producing thin, elastic sheets hardness = 2.5 specific gravity = 2.7	Muscovite
		white to greenish 1 direction of cleavage pearly luster hardness = I specific gravity = 2.8	Talc
Harder than glass	Absent	white, gray, pink, violet glassy luster conchoidal fracture hardness = 7 specific gravity = 2.6	Quartz
		olive green to yellow green glassy luster conchoidal fracture hardness = 7 specific gravity = 3.5	Olivine

Minerals...sigh!

Procedure

Examine the various unknown minerals, and note their characteristics on the following worksheet. Identify the different minerals by comparing your list to the mineral identification tables.

Separate metallic and nonmetallic minerals.

A. If the mineral is metallic, determine:
    1. streak
    2. color
    3. hardness
    4. any other distinguishing properties
    5. the name of the mineral

B. If the mineral is nonmetallic, determine:
    1. color (separate light minerals from dark minerals)
    2. hardness
    3. cleavage
    4. any other distinguishing properties
    5. the name of the mineral

Mineral Identification Sheet

Luster	Streak	Color	Hardness	Cleavage	Other Characteristics	Mineral Name

WHAT'S THAT MINERAL?

Summing Up

1. What distinguishing characteristic is used in identifying the following minerals?

   a. halite _____

   b. pyrite _____

   c. quartz _____

   d. biotite _____

   e. fluorite _____

   f. garnet _____

2. Indicate whether the following physical properties result from a mineral's crystalline structure or a mineral's chemical composition.

   a. crystal form _____

   b. color _____

   c. cleavage _____

   d. specific gravity _____

3. If a mineral does not exhibit a streak, is it metallic? Explain.

   _____

   _____

4. Which property is more reliable in mineral identification, color or streak? Why?

   _____

   _____

5. What physical properties distinguish biotite from muscovite?

   _____

   _____

6. What physical properties distinguish plagioclase feldspars from orthoclase feldspars?

   _____

   _____

CONCEPTUAL PHYSICAL SCIENCE	Activity

Rocks and Minerals

Rock Types

Rock Hunt

Purpose
To practice finding the geology all around you

Apparatus
desire

Discussion and Procedure

ACTIVITY 1
Go on a rock hunt. Now that you are more familiar with rocks and minerals, you can start your own collection. Gather rocks from the beach, an old river channel, a stream bed, a road cut, or even your back yard. Collect at least six different-looking rocks, and classify them into the three major rock groups. What telltale features help you classify the rocks? Can you identify your rocks by these features?

ACTIVITY 2
Rocks are found not only on the beach and in the mountains, but almost everywhere. In fact, if you live in a city you are probably surrounded by more rock materials than you realize. Take a field trip down any city street, and you will notice that most buildings are constructed from stone material. Many stone buildings are polished, providing an easy view of a rock's mineral composition, its texture, and hence its method of formation. Is the rock composed of visible crystals? Are the crystals interlocking? Are the  crystals flattened or deformed? What is the grain size? Are there fossils? Support your classification with your observations.

ACTIVITY 3
Observe the buildings in your locality. Look for the older buildings, most of which were built from local material. The different rock types used in the construction of some of these buildings can tell you much about local history. Is there a marble or granite quarry in your area? Try to trace the building material to its origin.

CONCEPTUAL PHYSICAL SCIENCE	Activity

Rocks and Minerals

What's That Rock?

Purpose
To identify rocks from the three different types: igneous, metamorphic, and sedimentary

Apparatus
collection of assorted igneous, sedimentary, and metamorphic rocks
dilute hydrochloric acid (HCl)

Discussion
The three major classes of rocks—igneous, sedimentary, and metamorphic—have their own distinct physical characteristics. By learning to identify the representative characteristics of each rock type, we may gain a better understanding of the history recorded in Earth's crust.

Procedure
You will be given three sets of rocks—igneous, sedimentary, and metamorphic. Your task is to identify the rocks in each set. The first set to identify consists of igneous rocks. Refer to Table 1, *Classification of Igneous Rocks*, to aid in your identification. The second set to identify consists of sedimentary rocks. For this part of the exercise refer to Table 2, *Classification of Clastic Sedimentary Rocks* and *Classification of Nonclastic Sedimentary Rocks*. To make identification and classification easier, Table 2 has been divided into two parts—Part A for clastic sedimentary rocks, and Part B for nonclastic sedimentary rocks. The final set to identify consists of metamorphic rocks. Refer to Table 3, *Classification of Metamorphic Rocks*, to help you distinguish among the different metamorphic rocks.

PART A: IDENTIFICATION OF IGNEOUS ROCKS
Molten magma welling up from within our planet produces igneous rock of two types—*intrusive* and *extrusive*. Intrusive rocks are formed from magma that solidified below Earth's surface, and extrusive rocks are formed from magma erupted at Earth's surface.

Step 1: The first step in identifying igneous rocks is to observe their texture. Texture is related to the cooling rate of magma in a rock's formation. Magma that solidifies below Earth's surface cools slowly, forming large, visible interlocking crystals that can be identified with the unaided eye. This coarse-grained texture is described as phaneritic. If the texture is exceptionally coarse-grained (visible minerals larger than your thumb), the rock is described as having a pegmatitic texture. Some intrusive rocks contain two distinctly different crystal sizes such that some minerals are conspicuously larger than the other minerals. This texture, with big crystals in a finer groundmass, is called porphyritic.

In contrast, magma that reaches the surface tends to cool rapidly, forming very fine-grained rocks. This fine-grained texture is described as aphanitic. When magma is cooled so quickly that there is not time for the atoms to form crystals, the texture is described as glassy. Many aphanitic rocks contain cavities left by gases escaping from the rapidly cooling magma. These gas bubble cavities are called vesicles, and the rocks that contain them are said to have a vesicular texture. If the gaseous magma cools very quickly, the texture that develops is described as frothy (foam-like glass). In a volcanic eruption, the forceful escape of gases causes rock fragments to be torn from the sides of the volcanic vent. These rock fragments combine with volcanic ash and cinders to produce a pyroclastic, or fragmental, texture.

Table 1

CLASSIFICATION OF IGNEOUS ROCKS				
	Composition			
	Light-Colored	Intermediate Color	Dark-Colored	Very Dark Color
	10–20% quartz K-feldspar > plagioclase ≈10% ferromagnesian minerals	No quartz plagioclase > K-feldspar 25–40% ferromagnesian minerals	No quartz plagioclase≈50% 50% ferromagnesian minerals	100% ferromagnesian minerals
Texture Pegmatitic (very coarse-grained)	Pegmatite			
Porphyritic (mixed crystal sizes)	Porphyry			
Phaneritic (coarse-grained)	Granite Granodiorite	Diorite	Gabbro	
Aphanitic (fine-grained)	Rhyolite Peridotite	Andesite	Basalt	Periodite Dunite
Glassy	Obsidian			
Porous (glassy, frothy)	Pumice		Scoria	
Pyroclastic (fragmental)	Volcanic tuff (fragments < 4 mm) Volcanic breccia (fragments > 4 mm)			

Step 2: Observe the color, and hence the chemical composition, of an igneous rock sample. Igneous rocks are either rich in silicon or poor in silicon. Silicon-rich (quartz-rich) rocks tend to be light in color, whereas silicon-poor rocks tend to be dark in color. Igneous rocks can therefore be divided into a light-colored group—quartz, feldspars, and muscovite; and a dark-colored group—the ferromagnesian minerals: biotite, pyroxene, hornblende, and olivine. The term *ferromagnesian* refers to minerals that contain iron (ferro) and magnesium (magnesian).

PART B: IDENTIFICATION OF SEDIMENTARY ROCKS

Rock material that has been weathered, subjected to erosion, and eventually consolidated into new rock is sedimentary rock. Sedimentary rocks are classified into two types—clastic and nonclastic. Clastic sedimentary rocks are formed from the compaction and cementation of fragmented rocks. Nonclastic sedimentary rocks are formed from the precipitation of minerals in a solution. This chemical process can occur directly, as a result of inorganic processes, or indirectly, as a result of biochemical reaction.

Step 1: Observe the texture of the rock. If the rock is composed of visible particle grains, the rock is probably clastic. Clastic sedimentary rocks are classified according to particle size. Large particles range from boulders to cobbles to pebbles. If the large particles are rounded, the rock is a conglomerate. If they are angular and uneven, the rock is a breccia. Medium-sized particles produce the various types of sandstone. Quartz sandstone is composed wholly of well-rounded and well-sorted quartz grains. Arkose is composed of quartz with about 25% feldspar grains. Graywacke is composed of both quartz and feldspar grains with fragments of broken rock in a clay-type matrix. Silts and clays consist of fine particles. Silt-sized particles produce siltstone, and clay-sized particles produce shale.

Step 2: If the rock does not fit into the clastic classification, the rock is probably nonclastic. Nonclastic sedimentary rocks are classified by their chemical composition. Inorganic processes produce the evaporites—minerals formed by the evaporation of saline water. Evaporites include rock salt and gypsum. Rock salt can be identified by its salty taste. Biochemical reactions produce carbonates by way of calcareous-secreting organisms, and chert by way of silica-secreting organisms. Limestone, chalk, and dolostone are carbonate rocks. Carbonate minerals react (fizz) to HCL. Chert can be identified by its waxy luster and conchoidal fracture. Coal (a biogenic sedimentary rock) is formed from the accumulation and compaction of vegetable matter.

Table 2

CLASSIFICATION OF CLASTIC SEDIMENTARY ROCKS			
Particle size		**Rock Name**	**Characteristics**
Coarse (> 2 mm)		Conglomerate	Rounded grain fragments
		Breccia	Angular grain fragments
Medium (1/16–2 mm)	Sandstone	Quartz	Quartz grains (often well-rounded, well-sorted)
		Arkose	Quartz and feldspar grains (often reddish color)
		Graywacke	Quartz grains, small rock fragments, and clay minerals (often grayish color)
Fine (1/256–1/16 mm)		Siltstone	Silt-sized particles, surface is slightly gritty
Fine (< 1/256 mm)		Shale	Clay-sized particles, surface has smooth feel, no grit

CLASSIFICATION OF NONCLASTIC SEDIMENTARY ROCKS			
Composition		**Rock Name**	**Characteristics**
Halite (NaCl)	Evaporites	Rock salt	Salty taste
Gypsum ($CaSO_4 \cdot 2H_2O$)		Gypsum	Inorganic precipitate
Organic matter		Coal	Compacted, carbonized plant remains
Calcium Carbonate ($CaCO_3$)		Limestone	Fine-grained, strong reaction to HCl
		Chalk	Microfossils; fine-grained
		Fossiliferous limestone	Macrofossils and fossil fragments
Dolomite ($CaMg(CO_3)_2$)		Dolostone	Reaction to HCl when in powdered form
Quartz (SiO_2)		Chert	Hard, dense, waxy luster, may show conchoidal fracture

WHAT'S THAT ROCK?

PART C: IDENTIFICATION OF METAMORPHIC ROCKS

Rock material that has been changed in form by high temperature or pressure is metamorphic rock. Metamorphic rocks are most easily classified and identified by their texture and (when a particular mineral is very obvious) by their mineralogy. Metamorphic rocks can be divided into two groups: foliated and nonfoliated.

Foliated metamorphic rocks have a directional texture and a layered appearance. The most common foliated metamorphic rocks are slate, phyllite, schist, and gneiss.

Slate is very fine-grained and is composed of minute mica flakes. The most noteworthy characteristic of slate is its excellent rock cleavage—the property by which a rock breaks into plate-like fragments along flat planes. Phyllite is composed of very fine crystals of either muscovite or chlorite that are larger than those in slate, but not large enough to be clearly identified. Phyllite is distinguished from slate by its glossy sheen. Schists have a very distinctive texture with a parallel arrangement of the sheet-structured minerals (mica, chlorite, and/or biotite). The minerals in a schist are often large enough to be easily identified with the naked eye. Because of this, schists are often named according to the major minerals in the rock (biotite schist, staurolite-garnet schist, and so on). Gneiss contains mostly granular, rather than platy, minerals. The most common minerals found in gneiss are quartz and feldspar. The foliation in this case is due to the segregation of light and dark minerals, rather than to alignment of platy minerals. Gneiss has a composition very similar to granite and is often derived from granite.

Nonfoliated rocks are mono-mineralic and thus lack any directional texture. Their texture can be described as coarsely crystalline. Common nonfoliated metamorphic rocks are marble and quartzite.

Table 3

CLASSIFICATION OF METAMORPHIC ROCKS			
Foliated Metamorphic Rock			
Crystal Size		**Rock Name**	**Characteristics**
Very fine, crystals not visible		Slate	Excellent rock cleavage
Fine grains, crystals not visible		Phyllite	Well-developed foliation; glossy sheen
Coarse texture Crystals visible with unaided eye Micaceous minerals Often contains large crystals	Schist	Muscovite schist	Mineral content reflects increasing metamorphism from top to bottom
		Chlorite schist	
		Biotite schist	
		Garnet schist	
		Staurolite schist	
		Kyanite schist	
		Sillimanite schist	
Coarse		Gneiss	Banding of light and dark minerals
Nonfoliated Metamorphic Rock			
Precursor Rock		**Rock Name**	**Characteristics**
Quartz sandstone		Quartzite	Interlocking quartz grains
Limestone		Marble	Interlocking calcite grains

Summing Up

1. Did you find evidence that igneous rocks can exhibit both fine and coarse textures? Describe the textures.

_____

_____

_____

_____

2. What factors promote the formation of large crystals? What types of rocks exhibit enlarged crystals?

_____

_____

_____

3. What distinguishing characteristics are exhibited in sedimentary rocks?

_____

_____

_____

4. How can we distinguish igneous rocks from metamorphic rocks?

_____

_____

_____

Going Further

For this exercise, you will first determine whether the rock is igneous, sedimentary, or metamorphic. Then you will use the classification tables for the different rock types to identify the rocks. Examine your rock specimen closely. Look at its texture. Can you see individual mineral grains? If so, the texture is coarse to medium-coarse. If the mineral grains are too small to be identified, the texture is fine. How are the grains arranged? Are the grains interlocking? Interlocking grains are formed by crystallization, so the rock is probably either igneous or metamorphic. Are the interlocking grains aligned? Do they show foliation? Foliation indicates metamorphism. Are the grains separated by irregular spaces filled with cementing material? Are fossils present? Does the rock react to acid? If so, the rock is sedimentary. If the rock is crystalline and shows no reaction to acid, check the hardness of the rock. Metamorphic and igneous rocks are harder than sedimentary rocks.

CONCEPTUAL PHYSICAL SCIENCE	**Activity**

Plate Tectonics and Earth's Interior   **Topographic Maps**

Top This

Purpose
To interpret topographic maps, to use points of known elevation to draw contour lines, and to use a topographic map to construct a topographic profile

Apparatus
topographic quadrangle map (provided by your instructor)
ruler, pencil, and eraser

Discussion
Maps provide a representation of Earth's surface and are a very useful tool. A map is a scaled-down, idealized representation of the real world. Everything on a map must be proportionally smaller than what it really is. Roads, waterways, mountains, ground area, and distance are all proportionally reduced in scale. A map's **scale** is defined as the relationship between distance on the map and distance on the ground. There are three ways to describe scale on a map. A *graphical scale* is a drawing, a line marked off with distance units. A *verbal scale* uses words to describe distance units—"one inch to one mile." And a *representative fraction* (rf) gives the proportion as a fraction or ratio, such as 1:24,000. This means that one unit of measure on the map—1 inch or 1 centimeter—is equal to 24,000 units of the same measure on the ground. If the scale is 1:10,000, then 1 inch on the map is equal to 10,000 inches on the ground. The first number refers to the map distance and is always 1. The second number refers to ground distance and changes depending on the scale.

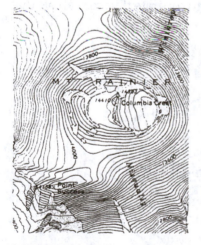

1:24,000 scale   1:100,000 scale   1:125,000 scale

Figure 1. U.S. Geological Survey Maps of the same area at different scales. The scale of a map tells us how much area is being represented and gives us the level of detail in that area. When comparing maps, we see that the smaller the scale, the larger the area represented, and the larger the scale, the smaller the area represented.

Large-Scale Map Example: 1:10,000	Small-Scale Map Example: 1:1,000,000
Shows a small area with a lot of detail. A large scale is good for urban, street, or hiking maps that require detail.	Shows a large area with very little detail. A small scale is good for world or regional maps that cover a large area.

SCALE CONVERSION

It is difficult to envision 10,000 inches, let alone 1,000,000 inches on the ground. In order to make these units more meaningful, we can convert them into miles or kilometers.

Question: One inch on a 1:25,000 scale represents what distance on the ground?

Answer Using conversion

1 foot = 12 inches
5280 feet = 1 mile
1 mile = 1.609 kilometers

$$25{,}000 \text{ in} \times \frac{1 \text{ ft}}{12 \text{ in}} \times \frac{1 \text{ mile}}{5280 \text{ ft}} = 0.394 \text{ mile}$$

$$0.394 \text{ mile} \times \frac{1.609 \text{ km}}{1 \text{ mile}} = 0.634 \text{ kilometers}$$

Question: One mile on the ground is represented by what distance (in inches) on a 1:25,000 scale map?

Answer $1 \text{ mi} \times \dfrac{5280 \text{ ft}}{1 \text{ mi}} \times \dfrac{12 \text{ in}}{1 \text{ ft}} \times \dfrac{1 \text{ in on map}}{25000 \text{ in}} = 2.53$ inches on map

RELIEF PORTRAYAL AND TOPOGRAPHIC MAPS

A **topographic map** is a two-dimensional representation of a three-dimensional land surface. Topographic maps show **relief**—the extent to which an area is flat or hilly. To show land surface form and vertical relief, topographic maps use contours. A **contour** line connects all points on the map having the same elevation above sea level. Each contour line acts to separate the areas above that elevation from the areas below it (Figures 2 and 3).

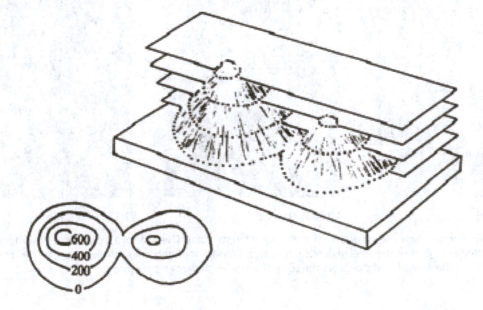

Figure 2

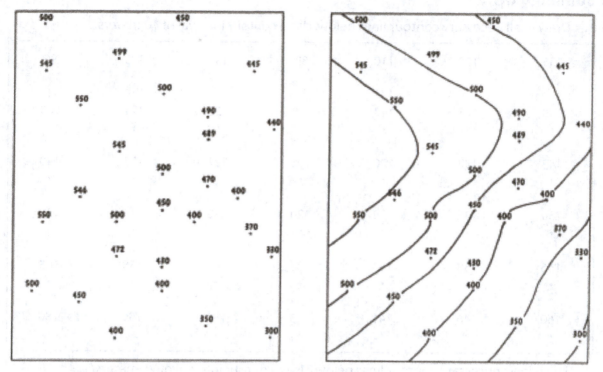

Figure 3. Known points of elevation are shown at the left. Contour lines from known points of elevation are shown at the right. The contour interval = 50 feet.

The vertical difference in elevation between contour lines is called the **contour interval**. The contour interval is usually stated at the lower margin of a topographic map. If it is not stated, look for a numbered contour line—an *index contour line* that is thicker and darker. Every fifth line is usually an index line. The choice of contour interval depends on the scale of the map.

All contour lines are multiples of the contour interval. For a contour interval of 10 feet, the sequence of contours would be 0, 10, 20, 30, 40, and so on. For a contour interval of 50 feet, the sequence of contours would be 0, 50, 100, 150, 200, and so on. If a point lies on a numbered contour line, the elevation is simply read from the line. If the contour line is unnumbered, the elevation can be found by counting the number of lines (contour intervals) above or below the index contour. Add or subtract this number from the index contour. If a point lies between contour lines, the elevation is approximated from the two contour lines it lies between. Elevations of specific points, such as mountain peaks, road crossings, or survey stations, can sometimes be read directly from the map. Most of these points are marked with a small "x" and a number for the elevation.

RULES FOR CONTOUR LINES
1. All points on a contour line must be of equal elevation.
2. Contour lines always close to form a closed path, like an irregular circle. If the contour line extends beyond the mapped area, the entire circle will not be seen.
3. Contour lines never cross one another.
4. Contour lines must never divide or split.
5. Closely spaced contour lines represent a steep slope; widely spaced contour lines represent a gentle slope.
6. A concentric series of closed contours represents a hill.
7. A concentric series of closed contours with hachure marks directed inward represents a closed depression.
8. Contour lines form a V pattern when crossing streams. The apex of the V always points upstream (uphill).

TOP THIS

Summing Up

1. Draw in all necessary contour lines below. Use a contour interval of 10 meters.

135	122	110	88	100	105	90
140	125	113	94	120	121	100
143	132	120	108	127	135	140
150	135	128	120	140	144	155
156	138	134	138	147	156	165
160	146	141	153	155	165	180

2. Draw in all necessary contour lines below. Use a contour interval of 25 meters.

165	124	75	63	123	187	198
132	100	99	67	71	114	148
116	148	169	102	60	63	67
115	176	197	148	100	69	54
101	147	206	187	149	113	70
93	123	200	203	148	87	159
82	98	154	205	185	193	175
60	101	166	169	156	153	199
56	95	164	132	141	160	173

3. Mark each contour line with its corresponding elevation. Use a contour interval of 40 feet.

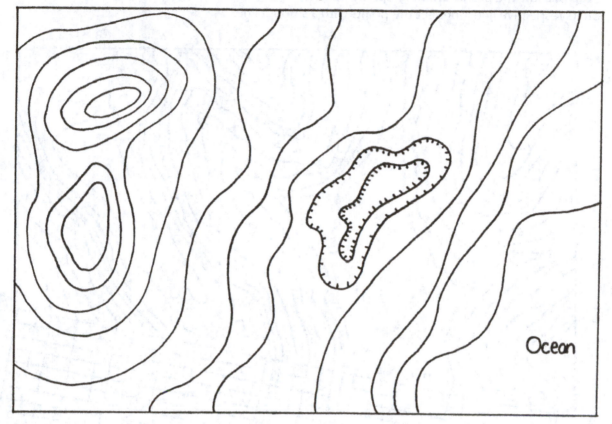

TOPOGRAPHIC PROFILE

A topographic map provides an overhead view of the landscape. A profile of the landscape across any area of the map can be constructed by drawing a straight line across the desired area. This profile is like a slice through the landscape affording us a side view. Figure 4 shows a method for constructing a topographic profile.

Procedure

Step 1: Mark a straight line across the map to indicate the line of profile; label it A–A'. Placement of the line can be anywhere, depending on what part of the landscape you want to view.

Step 2: Lay the edge of a sheet of paper along the line A–A'. Mark the position of each contour on the paper. Note the elevation for each mark. Other features, such as mountain crests and streams, should also be marked.

Step 3: Prepare a graph with the line A–A' as the horizontal axis and with elevation as the vertical axis. Each line will represent a contour line. All lines on the graph must be equally spaced. Label the lines so that the highest and lowest elevations along the line of profile fit into the graph. The vertical axis can be made to the same scale as the map, or, for a more impressive profile, the scale can be exaggerated. *Note:* The magnitude of any exaggeration must be stated for correct interpretation of the profile.

For vertical exaggeration, first decide on a vertical scale.

Example: 1 inch = 100 feet = 1200 inches (1:1200). To calculate the vertical exaggeration, divide the fractional vertical scale by the fractional horizontal scale.

Example: For a vertical scale of 1:10,000 and a horizontal scale of 1:50,000, the vertical exaggeration will be 5 times greater than the true relief.

TOP THIS

Step 4: Draw the profile on the graph by transferring the marked position of each contour elevation on its respective line. Connect the points with a smooth line.

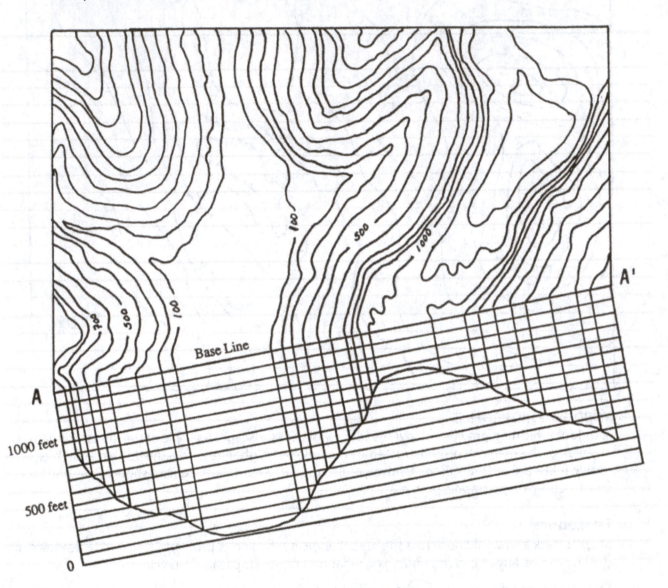

Figure 4. The profile of the terrain is plotted to an appropriate scale from the chosen base line.

CONCEPTUAL PHYSICAL SCIENCE	Activity

Plate Tectonics and Earth's Interior   **Geologic Cross Sections**

Over and Under

Purpose
To construct and interpret maps of geologic cross sections in the subsurface

Apparatus
protractor
compass
colored pencils and eraser

Discussion
Geologic cross sections show the three-dimensional structure of the subsurface. They provide a "pie slice" of the subsurface and are thus extremely useful for mineral, ore, and oil exploration. We can construct geologic cross sections by using surface information—outcrops of folded and faulted rock sequences, and igneous rock intrusions.

The study of rock deformation is called *structural geology*. When a rock is subjected to compressive stress, it begins to buckle and fold. If the stress overcomes the strength of the rock, the rock exhibits strain and breaks or faults. Structural geology interprets the different types of stress and strain.

We measure the orientation of deformed rock layers using strike and dip (Figure 1).

Strike is the trend or direction of a horizontal line in an inclined plane. The direction of strike is expressed relative to north. For example, "North X degrees west," or "North X degrees east."

Dip is the vertical angle between the horizontal plane and an inclined plane. Dip is always measured perpendicular to strike. We can think of the direction of dip as the direction a marble would roll down a plane. Hence, we express dip as (1) the direction a marble would roll, and (2) the angle between the inclined and horizontal planes.

Figure 1. Strike and dip. On the rock outcrop, the strike is the line formed from the intersection of a horizontal plane and the tilted rock strata. The dip is the angle between the horizontal and tilted strata (plane). The direction of dip is simply the geographical (N, S, E, or W) direction in which a marble would roll down the tilted plane. In the example shown, the outcrop is striking northwest and dipping 45° southwest.

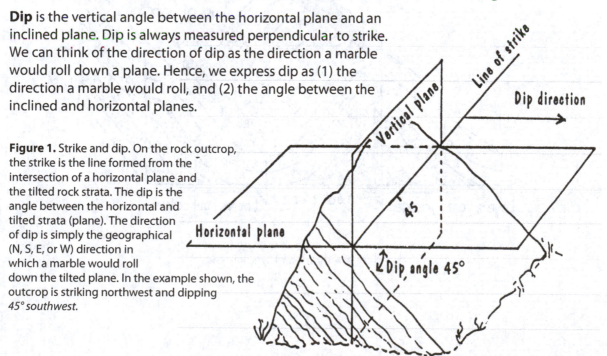

Geologic map symbols and symbols for strike and dip are shown in Figure 2.

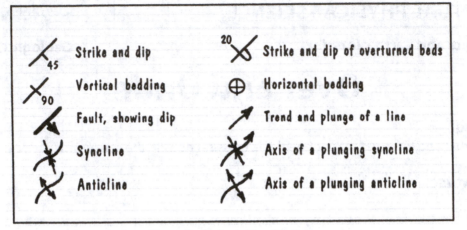

Figure 2. Geologic Map Symbols

The orientation of deformed layers can also be obtained through age relationships. Recall that sediments that settle out of water, such as in an ocean or a bay, are deposited in horizontal layers. The layer at the bottom was deposited first and is therefore the oldest in the sequence of layers. Each new layer is deposited on top of the previous layer. Therefore, in a sequence of sedimentary layers, the oldest layer is at the bottom of the sequence and the youngest layer is at the top of the sequence. As these sedimentary layers are subjected to stress, they fold and tilt. Each fold has an axis. If the tilted layers dip toward the fold axis, the fold is called a **syncline**. The rocks in the center, or core, of a syncline are younger than those away from the core. If the tilted layers dip away from the axis, the fold is called an **anticline**. The rocks in the core of the fold of an anticline are older than the rocks away from the core (Figure 3).

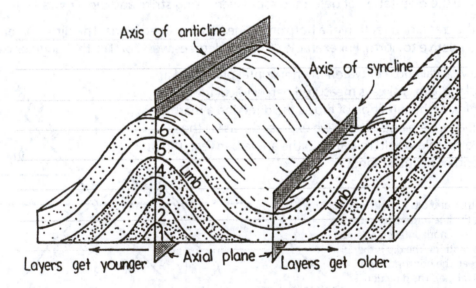

Figure 3. Anticline and Syncline Folds

The fold axis itself can be folded, but more often it is simply tilted. Folds in which the axis is tilted are known as *plunging folds* (Figure 4).

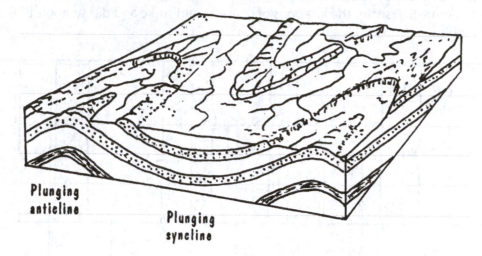

Figure 4. Plunging Folds

When stress exceeds the mechanical strength of a rock, the rock breaks or faults. The type of fault can be deciphered from the ages of the rocks on either side of the fault. Thrust and reverse faults are produced by compression (squeezing) forces. These types of faults have older, structurally deeper rocks pushed on top of younger, structurally higher rocks. Normal faults are produced by tensional (pulling) forces. This type of fault has younger, structurally higher rocks on top of older, structurally deeper rocks. The field evidence for normal faults is repeated layers. The evidence for faults at Earth's surface is "missing" or "repeated" layers of rock.

Shearing forces produce strike-slip faults where the rocks slide past one another. Strike slip faults show no vertical movement; the movement is horizontal. The different types of faults are illustrated in Figure 5.

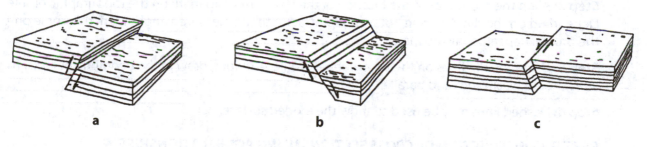

Figure 5
a Compressive forces produce thrust faults and reverse faults.
b Tensional forces produce normal faults.
c Shearing forces produce strike-slip faults.

Age sequence relationships can also be determined from faults and igneous intrusions. When igneous intrusions or faults cut through other rocks, the intrusions or faults must be younger than the rocks they cut.

In this exercise we will use age relationships, strike and dip orientations, and fault evidence to construct cross sections from map views. A map view shows only the surface features—rock type, sequence of rocks, strike and dip, and fault lines. The cross section is the subsurface extension of the surface view.

OVER AND UNDER

Procedure

PART A: CONSTRUCTION OF A CROSS SECTION USING STRIKE AND DIP MEASUREMENTS

Step 1: Refer to Figure 6. Transfer the locations of contacts and strike and dip symbols to the topographic profile.

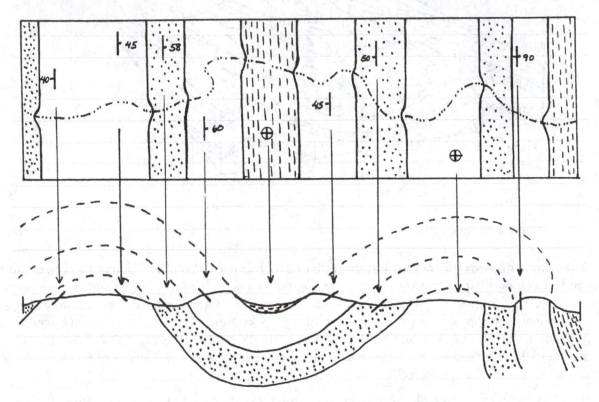

Figure 6

Step 2: Align the protractor to the location of the transferred dip mark on the topographic profile. Depending on the dip direction, rotate the protractor to the measured angle. Mark this angle onto the profile. Repeat for all dip marks.

Step 3: Using the dip lines on the topographic profile as guides, draw in the bedding contacts. The lines should be smooth and parallel.

Step 4: Dashed lines may be used to show the eroded surface.

PART B: CONSTRUCTION OF CROSS SECTIONS USING AGE RELATIONSHIPS

Step 1: Determine whether there is evidence of folding. Folding will show a mirror image about the fold axis of geologic layers.

Step 2: If the beds are folded, determine whether the bedding gets younger or older away from the axis of the fold.

Step 3: If the beds are not folded, the disturbed sequence may be due to faulting. Are there repeated layers? Are there missing rock layers? Are the layers offset?

PART C: THE CONSTRUCTION OF CROSS SECTIONS DISTURBED BY IGNEOUS INTRUSIONS

Step 1: Refer to Figure 7. If the beds are not folded, look to see whether there has been a disturbance—faulting or igneous intrusion. Faults and intrusions are always younger than the rock they cut into.

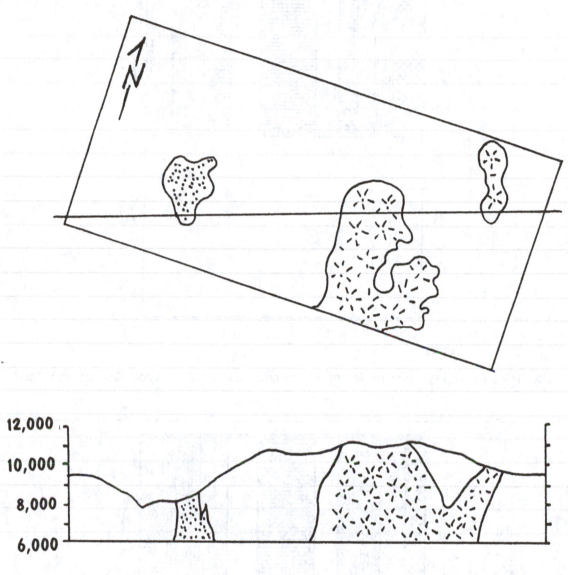

Figure 7

Step 2: Igneous intrusions work their way up from the deep subsurface. Because there is no way to tell the extent of an intrusion in the subsurface, there is a lot of interpretation.

OVER AND UNDER

Exercises

1. Use strike and dip symbols to determine the cross section. What is the structure?

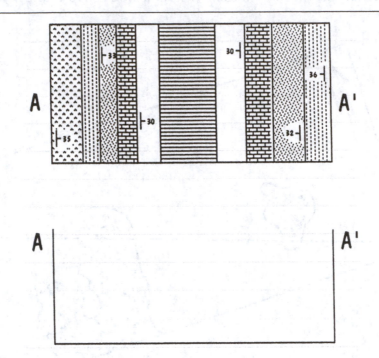

2. Use strike and dip symbols to determine the cross section. What type of structure is this?

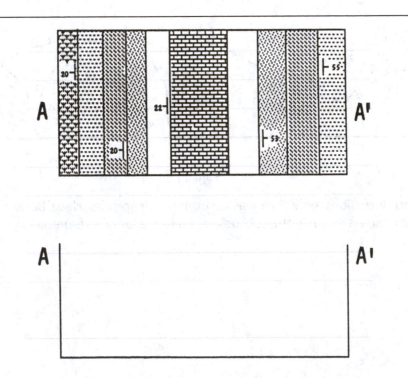

3. Use age relationships and general strike and dip symbols to determine the cross section. What type of structure is this?

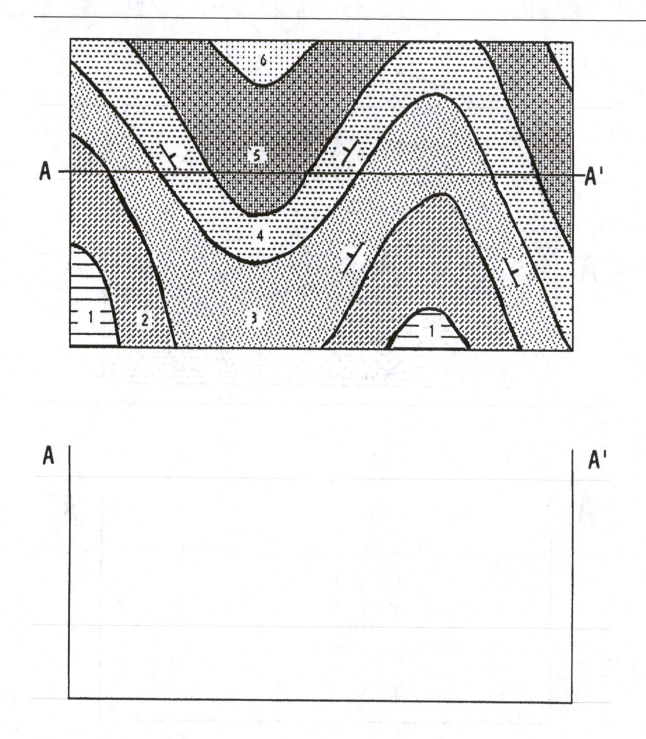

4. Use age relationships to determine the cross section. Note that 1 is the oldest layer and 6 is the youngest. What type of structure is this?

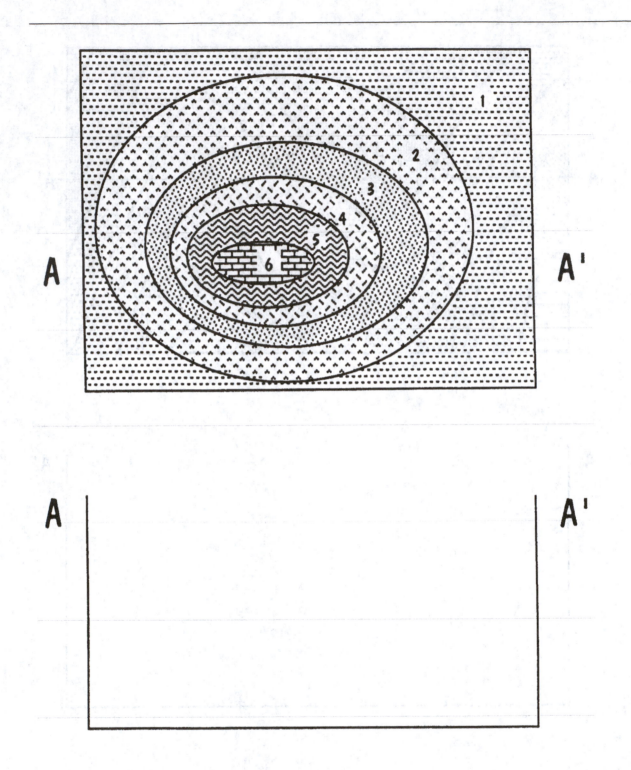

A   A'

A   A'

5. Use age relationships to determine the cross section. Note that 1 is the oldest layer and 5 is the youngest. What type of structure is this?

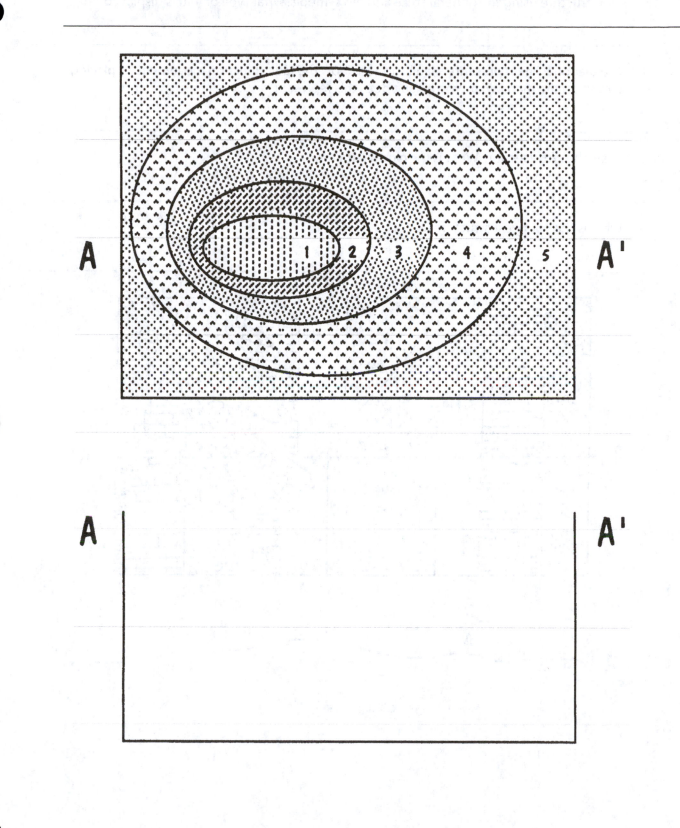

A   A'

6. This map shows evidence of disturbance and will require some interpretation. Determine the fold structure using the general strike and dip symbols. What type of fold is displayed?

_____

7. The diagonal line represents a fault surface. Show the fault direction movement by placing arrows on the map. What type of fault does this represent?

_____

8. Draw the cross section. What is the oldest structure?

_____

9. What is the youngest structure?

_____

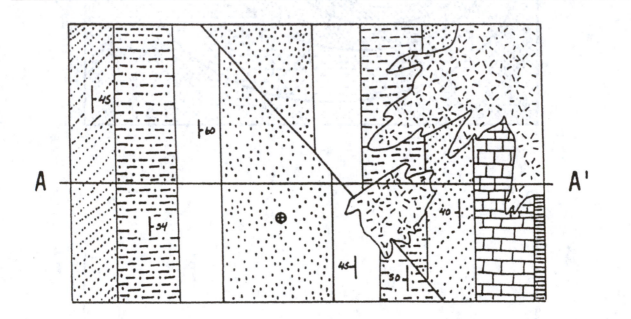

Summing Up

1. What type of structure is displayed in Exercise 1? Is the structure symmetrical or asymmetrical? Can the symmetry be determined from the map view?

_____

_____

_____

2. What type of structure is displayed in Exercise 2? Is the structure symmetrical or asymmetrical? Can the symmetry be determined from the map view?

_____

_____

_____

3. What type of structure is displayed in Exercise 3? What does the dip direction tell us about the structure?

_____

_____

_____

4. Exercise 4 and Exercise 5 are very similar to each other. How are they similar? How are they different? Mark the dip direction for both exercises.

_____

_____

_____

5. What type of fault is displayed in Exercise 6? What evidence supports your answer? What is the fold structure? Judging on the basis of the drawing, which structure is the oldest? Which is the youngest?

_____

_____

_____

CONCEPTUAL PHYSICAL SCIENCE | **Experiment**

Plate Tectonics and Earth's Interior | Volcanic Island Chain Formation

Hot Spot Volcanoes

Purpose
To investigate how tectonic plate motion is related to the formation of volcanic island chains

Apparatus
a clear plastic container such as a food storage box
a dropping bottle
red food coloring
hot water
cold water
a piece of Styrofoam or sponge to represent a tectonic plate

Discussion
Volcanoes form at diverging boundaries, where lava seeps out of fissures in the ocean floor and then collects over time. Volcanoes also form at convergent boundaries, where one oceanic plate subducts, partially melts, and then rises as magma and erupts. Sometimes, however, volcanoes develop in the middle of tectonic plates rather than at plate boundaries. In this case, the source of the volcano is a *hot spot*—a region deep within the mantle that is hot enough to melt the rock above it. The molten rock rises to the crust and erupts. Over time, the accumulated lava builds a volcano.

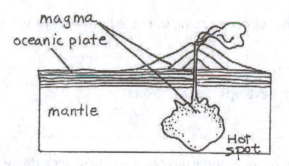

But tectonic plates move, as you know. What happens at Earth's surface when a tectonic plate shifts so that a volcano is no longer stationed over its source of magma?

Procedure
Step 1: Fill the container about 2/3 full with cold water.

Step 2: Fill the dropping bottle with hot tap water. Then add a few drops of red food coloring to it.

HOT SPOT VOLCANOES

Step 3: Place the open dropping bottle at the bottom center of the cold-water container as shown.

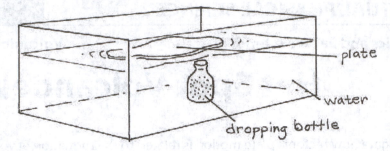

Step 4: Place your "tectonic plate" on the water's surface directly over the bottle. Observe the flow of the hot, colored water.

Step 5: Move your tectonic plate a short distance gently with one finger. Be careful not to disturb the water.

Step 6: Repeat Step 5 several more times until the edge of your tectonic plate lies over the dropping bottle.

Step 7: Remove your tectonic plate from the water. Observe the pattern of spots that the hot, colored water made on the plate.

Summing Up

1. What does the hot water represent?

_____

2. What does the cold water represent?

_____

3. What did the hot water escaping from the bottle do? How is that like magma from a hot spot?

_____

4. What landform develops over a hot spot?

_____

5. Draw a picture of your model tectonic plate here. Include the spots made by the drops of hot water. Can you tell in what direction the plate was moving by looking at the spots? If so, indicate this direction with an arrow on your drawing.

6. Draw a circle around the youngest volcano in your island chain.

7. In a volcanic chain formed inside plate boundaries, usually one volcano is active while the others are extinct. Which volcano is the active one on your model plate? Put a box around it on your drawing.

8. Look at the map of the Hawaiian Islands. The youngest volcano is Mauna Loa. It is at the southeastern end of the chain. Geologists believe all the islands in this chain formed over the same hot spot that lies under the Pacific Plate. Draw an arrow to represent the direction in which the Pacific Plate is moving.

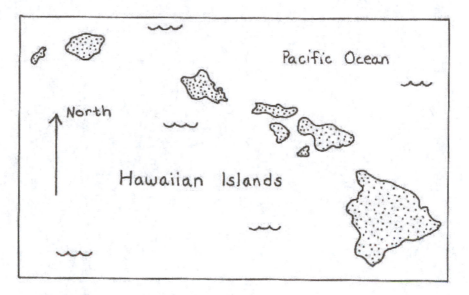

9. Only some volcanoes form over hot spots. Where else do volcanic island chains form? How do they form?

_____

_____

_____

HOT SPOT VOLCANOES

CONCEPTUAL PHYSICAL SCIENCE	Activity

Shaping Earth's Surface   **The Work of Groundwater**

Walking on Water

Purpose
To construct contour maps of the water table and determine flow paths

Apparatus
ruler
pencil and eraser
protractor

Discussion
Below the surface of continents is an extensive and accessible reservoir of fresh water. This reservoir of subsurface water is divided into two classes—soil moisture and groundwater. Groundwater is water that has percolated into the subsurface and saturated the open pore spaces in the rock or soil. The upper boundary of this saturated zone is called the **water table**. Water in the unsaturated zone above the water table is called *soil moisture*. The depth of the water table varies with precipitation and climate.

The water table tends to be a subdued version of the surface topography—sort of an underground surface. Water table contour maps, similar to surface contour maps, show the direction and speed of groundwater flow—information extremely useful for water supply management. For instance, simulated water table elevations generated by computer models can be compared with actual water table elevations in order to calibrate and verify the model. A groundwater model can reveal the best location for a well, and the impact that pumping from a new well will have on current water levels. Of particular interest today is the use of groundwater modeling to monitor contaminant transport in the subsurface.

Groundwater flow is influenced by gravitational force. Groundwater flows "downhill" underground, but the path it takes is dependent on *hydraulic head*, not on topography. Hydraulic head is higher where the water table is high, such as beneath a hill, and lower where the water table is low, such as beneath a stream valley. Responding to the force of gravity, water moves from high areas of the water table to low areas of the water table. As groundwater flows, it takes the path of least resistance—the shortest route. For any two lines of equal elevation, the shortest distance between them will be 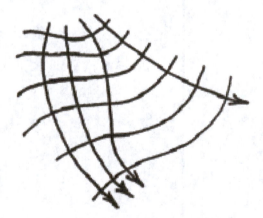 perpendicular to the lines. Thus, if we think of the path of groundwater flow as a *flow line*, the flow line will always be perpendicular to the contour lines.

PROBLEM 1: CONSTRUCTION OF WATER TABLE CONTOUR LINES

The construction of contour lines for a water table is very similar to the construction of surface topographic contour lines. The procedures and rules for surface contours also apply to water table contour maps. The only difference is that you are drawing contour lines of the water table below the ground's surface.

Draw contour lines for the elevation of the water table. Use a contour interval of 10 feet.

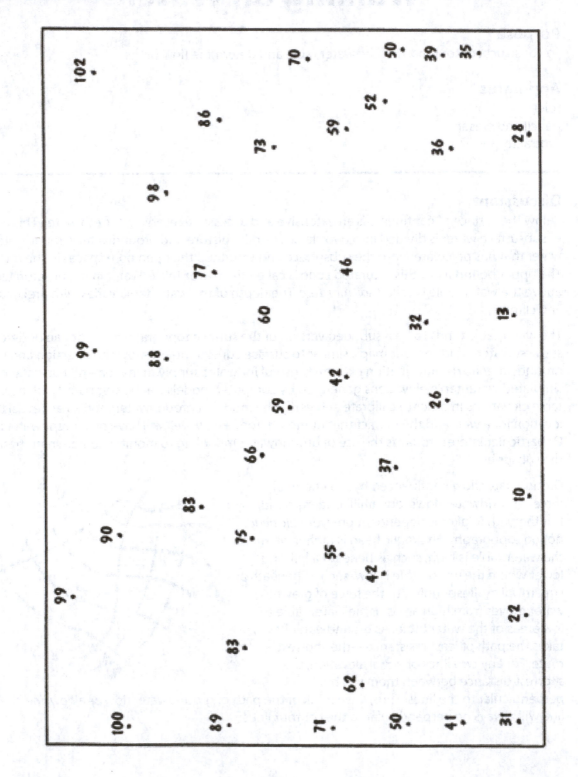

PROBLEM 2: CONSTRUCTION OF FLOW LINES

Water flows from areas of high hydraulic head to areas of lower hydraulic head. In mapping groundwater flow, the key concept is that the line of flow is perpendicular to the contour lines of the water table. You are provided with the cross section view as a guide to your thinking.

Draw flow lines on the water table contour maps, with arrows to show flow direction. Also draw arrows on the surface of the water table at cross sections to indicate the direction of flow.

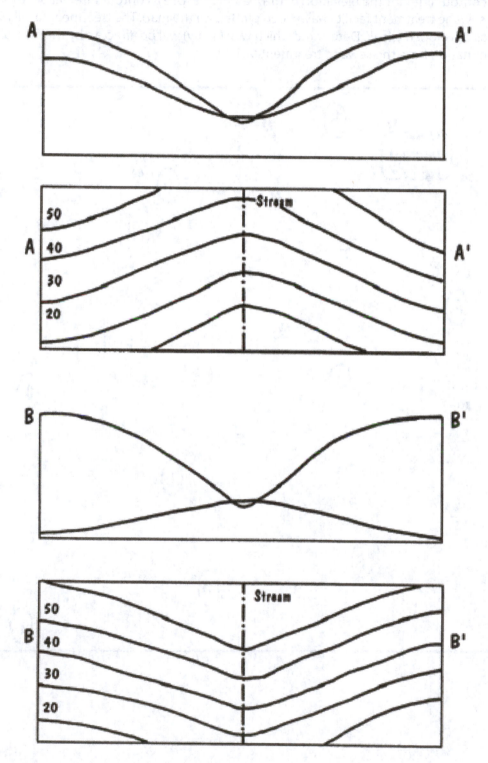

Knowledge of groundwater flow is important for solving problems of groundwater contamination. The most common groundwater contamination comes from sewage—drainage from septic tanks, inadequate or leaking sewer pipes, and farm waste areas. Contamination may also result from landfills and waste dumps where toxic materials and hazardous wastes leach down into the subsurface.

PROBLEM 3: CONTAMINANT FLOW MODEL

Draw contour lines for the elevation of the water table. Use a contour interval of 10 feet. Water wells and a sewage treatment facility have been plotted on the map. The treatment facility is in disrepair and has developed a leak. Determine which well (if any) will be affected by the flow of contamination from the sewage treatment facility.

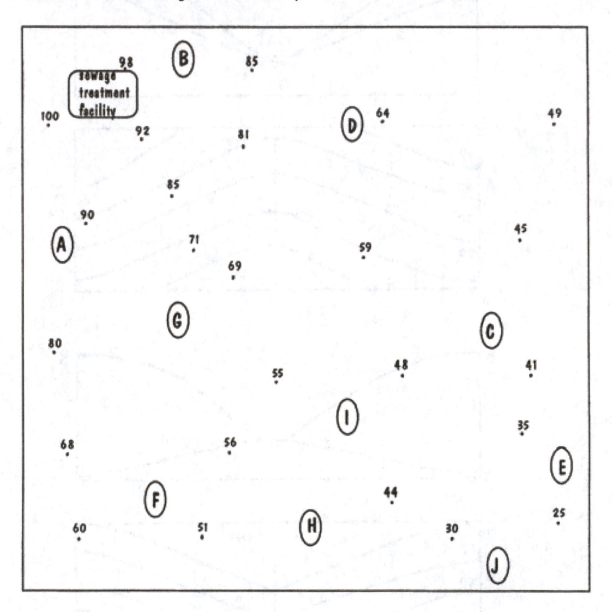

Summing Up

1. In Problem 2, what is the difference between stream A and stream B? In what type of climatic area (dry desert or wet forest) would we be likely to find stream A? In what type of climatic area would we be likely to find stream B? Explain.

    _____

    _____

    _____

2. In Problem 3, we are assuming equal rates of pumping for each well. If excessive pumping occurred at well H, would there be a change in the flow of contamination? Which well, if any, would probably become contaminated because of this pumping?

    _____

    _____

    _____

3. In Problem 3, we are assuming that all the soil in the subsurface is homogeneous (the same). If an impermeable clay lens were found at point 55, would the flow of sewage contamination be affected? What if an impermeable clay lens were found at point 35?

    _____

    _____

    _____

CONCEPTUAL PHYSICAL SCIENCE	Activity

Shaping Earth's Surface   **Erosion and Deposition**

Stream Study

Purpose
To study the way in which moving water erodes and then deposits sediments of different sizes

Apparatus
stream table
fine sand
coarse sand
pebbles
bucket
large spoon or scoop

Discussion
Streams reshape Earth's surface in dramatic ways. You may know from reading the text that the Grand Canyon is a spectacular example of stream erosion. And you might have seen pictures of powerful floods that have washed away homes. But even when stream erosion isn't obvious, even when streams flow peacefully in their banks, they are steadily reworking the land.

What exactly do streams do to Earth's surface as they move across it? In this experiment, you will use a physical model of a stream—*a stream table*—to study the ways in which streams alter the landscape. And you will learn some things about river erosion and deposition that could affect where you choose to live.

Procedure
Step 1: Mix the sand and pebbles in the bucket. Scoop the mixture to a depth of 5 cm in the stream table.

Step 2: Wet the sand mixture with a small amount of water. Pat the sand mixture into a gentle slope.

Step 3: Run your finger along the sand in a zig-zag pattern to make a slightly curvy path that resembles a stream bed.

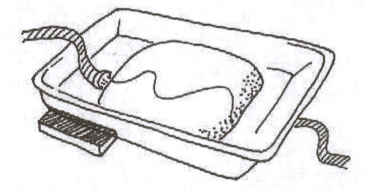

Step 4: Adjust your water supply so that it flows slowly into the tray. Allow the water to run in the stream table for 20 minutes. While you watch the water flow, answer the following questions:

1. **Where does the water slow down?**

_____

2. **Where does the water speed up?**

_____

3. **Where do you see erosion occurring?**

_____

4. **Where do you see deposition occurring?**

_____

Summing Up

1. Look carefully at the material deposited at different places. What pattern do you see?

_____

_____

2. Is the speed of water flow in a stream related to the size of the sediments that are deposited?

_____

3. If you answered "yes" to the previous question, what is the relationship between stream speed and the size of deposited sediments?

_____

_____

4. Look at the illustration below. Using evidence gathered in your stream table experiment, answer the questions below.

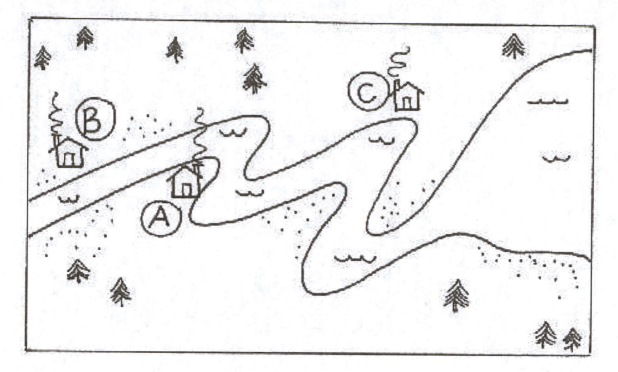

a. Which house has been built in the most desirable location? Why do you think so?

_____

_____

b. Which house may get a bigger front yard over time? Why do you think so?

_____

_____

c. Which house is built in the least desirable location? Explain your thinking.

_____

_____

CONCEPTUAL PHYSICAL SCIENCE	**Experiment**

Shaping Earth's Surface

Movement of Water Through Soil

Soil and Water

Purpose
To investigate the movement of water through different types of soil

Apparatus
3 soil samples: topsoil, clay, and sand
water
stopwatch (or clock with second hand)
4 500-mL beakers (or equivalent transparent containers)

Discussion
Rainwater that doesn't evaporate or form runoff moves downward into the tiny spaces between soil particles and rock underground. The small spaces that exist between soil or rock particles are called *pore spaces*. The amount of water that a soil can hold in its pore spaces is its *porosity*.

Why do we care about porosity? One reason is that the porosity of a soil determines how well plants can grow in it. You can measure porosity by measuring how much water a soil sample holds.

Some soils and regions of subsurface rock have pores that are well connected. Water flows through them easily. In this case, we say the soil or rock is *permeable*. Wells are drilled into permeable subsurface rock because groundwater can be most easily obtained there. You can measure the permeability of soil or rock by measuring how long it takes water to run through it.

In this activity, you will determine the porosity and permeability of several common kinds of soil.

Procedure
Step 1. Fill each container with a different kind of soil. Fill each container up to the same level (about 3 inches deep). Do not pack the soil. Label each container with the type of soil it contains.

Step 2. Before pouring any water onto the soil, make the following predictions.
a. Predict: Which soil is most permeable? Why do you think so?

_____

_____

b. Predict: Which soil is most porous? Why do you think so?

_____

_____

Step 3. Pour 100 mL of water into your beaker. *Very slowly*, pour the water into one of your soil samples. Pour the water in the middle of the soil rather than along the side of its container. Stop pouring when water flows out the bottom of the soil. Have a partner time how long it takes the water to reach the bottom of the container. Record the time for each sample in the following data table.

Table

Type of Soil	Volume of Water Absorbed by the Soil	Time for Water to Pass through the Soil

Summing Up

1. Which soil had the greatest permeability?

_____

2. Which soil was the least permeable?

_____

3. Which soil was the most porous?

_____

4. Which soil was the least porous?

_____

5. How is the size of the particles in the soil related to its permeability?

_____

_____

6. Which soil would allow the most runoff in a rainstorm? Defend your answer.

_____

_____

7. Some varieties of roses and grapes grow well in sandy soil. Why?

_____

_____

8. Soil that allows water to drain quickly is best for building roads and buildings upon. Which type of soil that you tested would be the best choice to build on?

_____

_____

9. The majority of plants grow best in soil that holds water for a time. Based on your observations, which soil type is the best choice for your garden?

_____

_____

CONCEPTUAL PHYSICAL SCIENCE	Activity

Geologic Time Scale and Relative Dating

Purpose

To use the principles of relative dating and the geologic time scale to investigate the relative dating of rock layers

Apparatus

Conceptual Physical Science text, pencil, and eraser

Discussion

Earth's history is recorded in the rocks of its crust. We cannot see the processes that created rocks that are millions and even billions of years old, but we can observe present-day rock-forming processes and their results. If the resulting present-day characteristics are similar to the characteristics of the older rocks in question, we can infer that the processes that operated in the past are the same as those that operate today. This is the *Principle of Uniformitarianism*—the physical, chemical, and biological laws that operated in the distant past continue to operate today. Simply stated, the present is the key to the past.

Geologists of long ago observed how present-day sediment is deposited horizontally, one layer at a time. This basic observation, called **original horizontality**, is one of the fundamental principles of **relative dating**—the process of determining the order in which geologic events occurred. The principles of relative dating, as the name suggests, help us determine the chronological order of geologic events, but not the actual dates of these events.

There are five key principles of relative dating:

Original horizontality

Sediment layers are deposited evenly, with each new layer laid down nearly horizontally over older rocks and sediment.

Superposition

In an undeformed, layered sequence of rocks, each layer is older than the one above and younger than the one below. In other words, the rock record was formed from the bottom layer to the top.

Cross-cutting relationships

Any type of igneous intrusion or fault that cuts through a rock body or layer is younger than the rock through which it cuts.

Inclusions

Any rock inclusion is older than the rock containing it.

Faunal succession

The evolution of life is recorded in the rock record in the form of fossils. Fossil organisms follow one another in a definite, irreversible sequence. Once an organism becomes extinct, it is never again seen at later times in the fossil record.

PROBLEM 1: RELATIVE TIME—*WHAT CAME FIRST?*

The cross section below shows the results of several geologic events. To the right of the diagram, list the sequences of geologic history starting at the bottom with the oldest event, and ending with the youngest event (at the top).

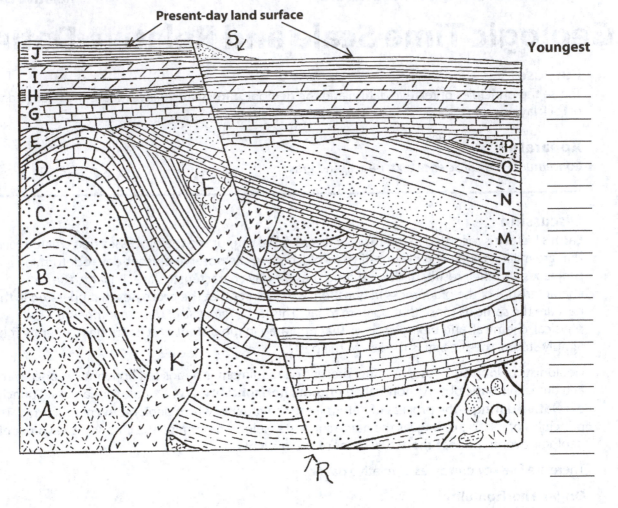

Present-day land surface

Youngest

_____
_____
_____
_____
_____
_____
_____
_____
_____
_____
_____
_____
_____
_____
_____
_____
_____
_____

Oldest

Summing Up

1. What geologic events happened before layer "L" was deposited? [*Hint*: Think tectonics.]

_____

_____

_____

_____

2. What type of event does feature "Q" represent? Which principles help you to approximate the age range of "Q"?

_____

_____

_____

_____

3. Feature "R" is a fault. What is the approximate age range of "R"? What type of fault is "R"? What type of force generated the fault, compression or tension? Support your answer.

_____

_____

_____

_____

PROBLEM 2: LIFE AND THE GEOLOGIC TIME SCALE

The geologic time scale was developed through the use of relative dating, and fossils played a large role. Through meticulous analysis of fossils, geologists were able to arrange different groups of fossils—and the rock layers in which they were found—into a chronological sequence.

Because each time period can be recognized by the fossils it contains, the fossils found in rocks can be used to identify other rocks of the same age in other regions of Earth. Actual specific dates were later applied to the geologic time scale via radiometric dating (a process that measures the ratio of radioactive isotopes to their decay product).

Refer to your text to match each life form with the appropriate geologic time period.

Geologic Time Period

Quaternary _____

Tertiary _____

Cretaceous _____

Jurassic _____

Triassic _____

Permian _____

Carboniferous _____

Devonian _____

Silurian _____

Ordovician _____

Cambrian _____

Life Form

A) Trilobite

B) Flowering plants

C) Age of humans

D) Swampy environments

E) Emergence of dinosaurs

F) Age of mammals

G) True pines and redwoods

H) First reptiles

I) Age of fishes

J) Emergence of land plants

K) First fish

CONCEPTUAL PHYSICAL SCIENCE	Activity

Reading the Rock Record

Purpose

The *principle of superposition* can be used to determine the relative ages of rock layers at a single location. Such locations, where rocks are exposed at Earth's surface, are called *outcrops*. For regions with several outcrops, geologists also use the *principle of lateral continuity* to determine relative ages. The principle of lateral continuity states that the layer-forming sediments we now see in rock outcrops were originally deposited and solidified into contiguous horizontal layers. When fully formed, such layers extended continuously for great distances until an obstruction was encountered. In this activity, you will reconstruct a simple geologic history of a hypothetical area by examining diagrams of several rock outcrops from that area. You will then use fossils and their respective periods of existence in geologic time to determine the ages of the rock layers.

Apparatus

Conceptual Physical Science text, pencil, and eraser

Discussion

Perhaps the world's most spectacular display of a rock record is the Grand Canyon of the Colorado River in Arizona. The many layers of sedimentary rock exposed in the canyon walls and the thicknesses of the different rock layers are testimony to great geologic activity over millions of years. The vertical sequence of its cake-like layering provides a great example for understanding relative dating. And the roughly horizontal rock layers that often stretch continuously for hundreds of miles are wonderful examples of lateral continuity.

But even the Grand Canyon does not exhibit a complete and continuous geologic history. In fact, even though some rock layers may have been deposited without interruption, nowhere on Earth do we find a continuous sequence of rock from Earth's formation to the present time. Weathering and erosion, crustal uplifts, and other geologic processes interrupt and/or prevent the deposition of a continuous sequence. These time gaps, or breaks, in the rock record are called **unconformities**. An unconformity between two rock layers can represent one or more rock layers that used to be present but were eroded away before deposition of the overlying layer. Or an unconformity can represent a long period of time in which deposition did not occur. Thus an unconformity indicates a significant time gap between the underlying and overlying layers. Unconformities are discovered by carefully observing the relationships of rock layers and fossils.

Correlation of rock outcrops

Piecing together Earth's history and conducting other geologic investigations often involves matching up, or correlating, rock outcrops that are separated geographically by large distances. In contrast to the Grand Canyon, most rock layers are not horizontally continuous from one rock outcrop to the next. The processes that contribute to unconformities— faulting, folding, uplift, and weathering and erosion—also act to break the lateral continuity of layers. And younger, unconsolidated sediment and soil end up burying older rock layers so that all that remains visible are geographically separated rock outcrops. Because of such occurrences, it is often difficult to tell, within a study area, whether similar-looking rock layers at different outcrops used to be part of the same, once-continuous layer.

To correlate one isolated rock outcrop with another involves looking for similar features among the different outcrops. We begin with the physical characteristics of the rocks—color, mineralogy, and grain size, for example. If the various rock layers in the isolated outcrops have characteristics that are similar enough, and they have consistent vertical sequences, we can safely assume that the layers were at one time continuous.

For example, in outcrop A (Figure 1) we might have a limestone layer sandwiched between a black shale layer on top and a red sandstone layer on the bottom. Then, if we find a similar sequence of rock layers at outcrops B and C, the three geographically separated outcrops are correlated and were once laterally continuous layers.

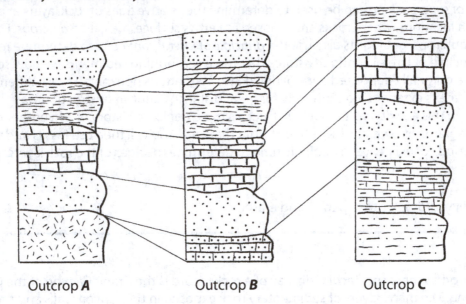

Outcrop **A**   Outcrop **B**   Outcrop **C**

Figure 1. Correlation of outcrops A, B, and C shows that the sandstone, limestone, and shale layers are in the same relative vertical sequence—evidence that all three outcrops were, at one time, laterally continuous. Parallel lines are drawn between the outcrops for the layers on which the correlation is based. Note the "pinched" lines from outcrop B to outcrop A. These "pinched" lines show that two rock layers are missing from outcrop A. The boundaries between the two uppermost and the two lowermost rock layers in outcrop A are unconformities. The missing layers indicate two time gaps in the rock record at outcrop A.

The importance of fossils

The job of correlating rock layers between outcrops is easier when fossils are found. Recall the principle of faunal succession—that the evolution of life is recorded in the form of fossils, and that fossil organisms follow one another in a definite, irreversible sequence in time. When fossils are found, they can be used as distinctive markers in sequences of sedimentary rocks. They not only aid in correlating rock layers, they can also be used as age indicators of the rock.

A group of fossil organisms that lived during the same range of time and are found in one or more layers is called a **fossil assemblage**. So if our limestone layer at outcrop A contains an assemblage of certain marine-type fossils, and then the same assemblage is found at outcrops B and C, we have even stronger evidence to support what our correlation suggests: that the rock layers were at one time continuous. Further investigation into the types of fossils in the assemblages enables us to determine the time period(s) during which the rocks were formed.

It is important to note that the presence of a similar or even identical fossil assemblage from rocks in two different locations does not mean that those rocks are exactly the same age. Correlation of fossil assemblages tells us that the ages of rock layers at different outcrops are similar in that they were formed within the same time range. Even though there may be an uncertainty of thousands to millions of years, such an approximation is sufficient for many purposes. (Remember, Earth is 4.5 *billion* years old!)

Construction of a stratigraphic column

The first geologic time scale was constructed by using the principle of superposition and the correlation of fossil assemblages in sedimentary rocks from numerous regions. Primitive fossils were placed at the bottom of the time scale column not because of their primitive nature, but because of superposition—they were found at the lowest layers in several (or many) different sequences of rock layers. The assumption was that rocks—and the fossils in them—found at the bottom of a sequence of rock layers are older than those found at the top.

The following exercise will provide insight into the methods that were used to construct the geologic time scale. Geologists show the vertical chronological succession of sedimentary rock layers from oldest at the bottom to youngest at the top by first constructing a diagram called a **stratigraphic column**.

Procedure

Four stratigraphic columns are depicted in Figure 2. The columnar stratigraphic sections are from outcrops in four widely separated, hypothetical regions. The rock layers in these outcrops were at one time continuous among all four regions. Various episodes of uplift and erosion have broken the continuity these rock layers used to share.

Please note: P *is the oldest layer.*

Step 1: To see which layers match up, draw lines connecting the boundaries of similar rock units from one outcrop to the next (see the example in Figure 1). In the individual stratigraphic columns, which rock layers are missing between the various regions? Note the missing layers for each column.

Step 2: In the blank column to the left, construct a new stratigraphic column based on the correlation among the four different regions. The new column will include every rock layer found at the four outcrops. Draw the different rock layers, or write the letter for each layer, in their proper order—oldest at the bottom to youngest at the top. Once again, remember that P is the oldest layer.

Going Further

Fossils document the evolution of life through time. Certain fossils that are geographically widespread and limited to a short span of geologic time are called index fossils. Index fossils can be used for correlating rocks of the same age and determining the approximate ages of rock layers. When a specific index fossil cannot be found, geologists look for groups of fossils—a fossil assemblage. In many ways, an assemblage of fossils can provide even more precise information. The more data one has, the clearer the picture becomes.

Now back to our hypothetical rock outcrops. The different layers of rock contain different fossil assemblages. Your task now is to assign time periods to the layers.

Step 3: Figure 3 shows the age ranges for fossils in our hypothetical study area. Many fossils overlap one another in time. Given that all rock units are for the Paleozoic Era, assign time periods to the rock layers in Figure 2. Write the names of the periods next to the respective rock layers in your stratigraphic column from Step 2.

READING THE ROCK RECORD

Summing Up

1. List the missing rock layers for each region.

_____

_____

2. The appearance of identical fossils in each of the sections facilitates the correlation of rock layers between outcrops. The reptile in three of the outcrops is not present in Region 1. What could this indicate? Support your answer.

_____

_____

3. Our particular fish fossil is found in layers Z and X and has a time range extending from the Devonian to the Permian. How can we assign layer X a specific time period? Explain.

_____

_____

4. There are several rock layers that have more than one designated time period. Name these layers. How can fossils extending over more than one time period be found in one rock layer?

_____

_____

5. There are many different types of fossils in our hypothetical rock layers. What do the majority of these fossils have in common? (Need a hint? Think habitat and environment.)

_____

_____

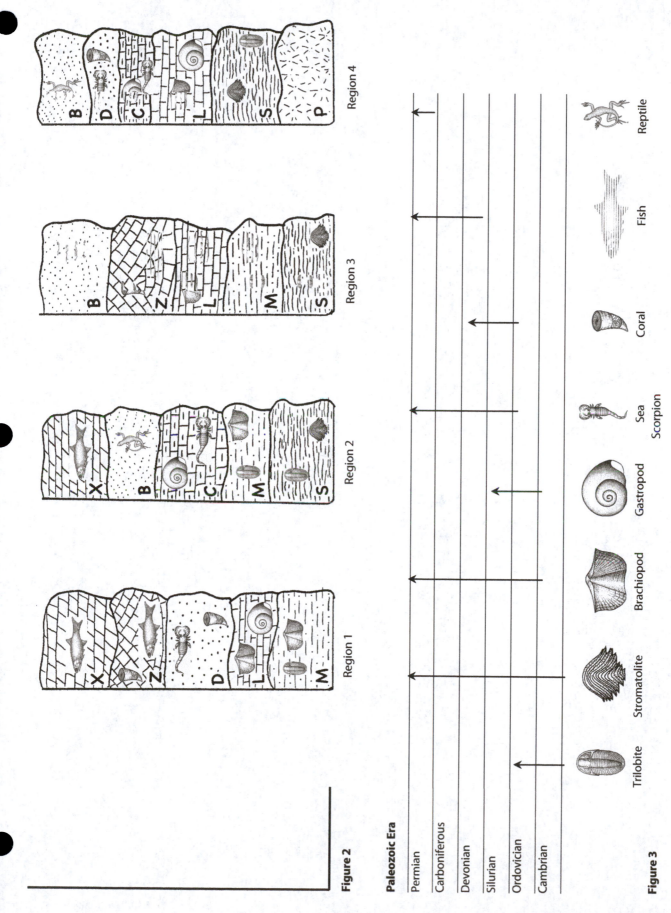

Figure 2

Region 1   Region 2   Region 3   Region 4

Paleozoic Era

Permian
Carboniferous
Devonian
Silurian
Ordovician
Cambrian

Trilobite   Stromatolite   Brachiopod   Gastropod   Sea Scorpion   Coral   Fish   Reptile

Figure 3

READING THE ROCK RECORD

CONCEPTUAL PHYSICAL SCIENCE | **Experiment**

The Oceans, Atmosphere, and Climatic Effects | The Power of the Sun

Solar Power I

Purpose

To measure the Sun's power output by comparison with the power output of a 100-watt light bulb

Apparatus

2 cm × 6 cm piece of aluminum foil with one side thinly coated with flat black paint
clear tape
glass jar with a hole in the metal lid
fine-scaled thermometer

meterstick
one-hole stopper to fit the hole in the jar lid
clear 100-watt light bulb with receptacle

Discussion

To measure the power output of a light bulb is nothing to write home about—but to measure the power output of the Sun using only household equipment is a different story. In this experiment, we will do just that—measure to a fair approximation the power output of the Sun. We will do this by comparing the light from a bulb of known wattage with the light from the Sun, using the simple ratio

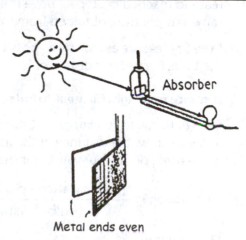

$$\frac{\text{Sun's wattage}}{\text{Sun's distance}^2} = \frac{\text{bulb's wattage}}{\text{bulb's distance}^2}$$

We **will** need a sunny day to do this experiment.

Metal ends even

Figure 1

Procedure

Step 1: With the blackened side facing out, fold the middle of the foil strip around the thermometer bulb, as shown in Figure 1. The ends of the metal strip should line up evenly.

Step 2: Crimp the foil so that it completely surrounds the bulb, as in Figure 2. Bend each end of the foil strip outward (Figure 3). On a tabletop or with a meterstick, make a flat, even surface. Use a piece of clear tape to hold the foil to the thermometer.

Crimp here

Figure 2

Step 3: Insert the free end of the thermometer into the one hole-stopper (soapy water or glycerin helps). Remove the lid from the jar, and place the stopper in the lid from the bottom side. Slide the thermometer until the foil strip is located in the middle of the jar. Place the lid on the jar.

Step 4: Position the jar indoors near a window so that the Sun will shine on it. Prop it at an angle so that the blackened side of the foil strip is perpendicular to the rays of the Sun. Keep it in this position in the sunlight until a stable temperature is reached. If you prefer, do this outside, which means doing Steps 5 and 6 outside.

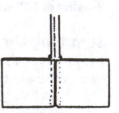

Bend blades outward

Figure 3

Stable temperature = _____ °C

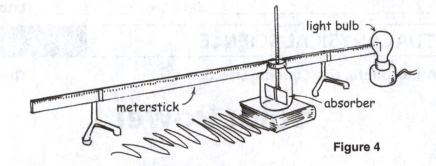

Figure 4

Step 5: Now find the conditions for bringing the foil to the same temperature with a clear 100-watt light bulb. Set the meterstick on the table. Place the clear 100-watt light bulb with its filament located at the 0-cm mark of the meterstick (Figure 4). Center the jar at the 95-cm mark with the blackened side of the foil strip perpendicular to the light rays from the bulb. You may need to put some books under the jar to align it properly.

Step 6: Turn the light bulb on. Slowly move the jar toward the light bulb, 5 centimeters at a time, allowing the thermometer temperature to stabilize each time. As the temperature approaches the reading reached in Step 4, move the jar only 1 centimeter at a time. When the bulb maintains the same temperature obtained from the Sun for about two minutes, turn the light bulb off.

Step 7: Measure as exactly as possible the distance in meters between the foil and the filament of the bulb. Record this distance.

Distance from light filament to foil strip = _____ m

Step 8: Because you know that the Sun's distance from the thermometer in meters is 1.5×10^{11} m, and you know the distance of the light bulb from the thermometer and the wattage of the bulb, you know three of the four values for the ratio equality stated earlier. With simple rearrangement,

$$\text{Sun's wattage} = \frac{(\text{bulb's wattage})(\text{Sun's distance})^2}{(\text{bulb's distance})^2}$$

Show your work:

Sun's wattage = _____ W

Step 9: Use the Sun's wattage to compute the number of 100-watt light bulbs needed to equal the Sun's power. Show your work.

Number of 100-watt light bulbs needed = _____

Summing Up

The accepted value for the Sun's power is 3.8×10^{26} W. How close was your experimental value? List factors you can think of that contribute to the difference.

_____

_____

CONCEPTUAL PHYSICAL SCIENCE

Experiment

The Oceans, Atmosphere, and Climatic Effects

Solar Energy

Solar Power II

Purpose

To measure the amount of solar energy per minute that reaches Earth's surface, and from this to estimate the Sun's power output

Apparatus

2 Styrofoam cups
water
plastic wrap
thermometer (Celsius)

graduated cylinder
blue and green food coloring
rubber band
meterstick

Discussion

How do we know how much energy the Sun radiates? First, we assume that the Sun radiates energy equally in all directions. Imagine a heat detector so big that it completely surrounds the Sun, like an enormous basketball with the Sun at its center. Then, the heat energy reaching this detector each second would be the Sun's power output. Or if our detector were half a basketball and caught half the Sun's energy, then we would multiply the detector reading by 2 to compute the total solar output. If our detector were one-quarter of a basketball, and caught one-quarter of the Sun's energy, then its total output would be 4 times the detector reading.

Now that you have the concept, suppose that our detector is the water surface area of a full Styrofoam cup here on Earth facing the Sun. Then, that area is only a tiny fraction of the area surrounding the Sun. If we figure what that fraction is, and also measure the energy captured by the water in our cup, we can tell how much total energy the Sun radiates. In this experiment, we will measure the amount of solar energy that reaches a Styrofoam cup and relate it to the amount of solar energy that falls on Earth. We **will** need a sunny day for this.

Procedure

Step 1: Using a graduated cylinder, measure and record the amount of water needed to fill a Styrofoam cup. Add a small amount of blue and green food coloring to make the water dark (and a better absorber of solar energy).

Volume of water = _____ mL

Mass of water = _____ g

Nest the water-filled cup in a second Styrofoam cup (for better insulation).

Step 2: Measure the water temperature and record it.

Initial water temperature = _____ °C

Step 3: Cover the doubled cup with plastic wrap, and seal it with a rubber band to prevent water leaks. Place the cup in the sunlight, tipped if necessary so that the plastic surface is approximately perpendicular to the Sun's rays. Let it warm by sunlight for 10 minutes.

Step 4: Remove the plastic wrap. Stir the water in the cup gently with the thermometer. Measure the final water temperature and record it. Find the difference in water temperature before and after the cup has been in the sunlight.

THE QUANTITY Q OF HEAT ENERGY COLLECTED BY THE WATER = MASS OF THE WATER × ITS SPECIFIC HEAT (c = 1 °C/g) × ΔT, ITS CHANGE IN TEMPERATURE

Final water temperature = _____ °C

Temperature difference = _____ °C

Step 5: Compute the surface area of the top of the cup in square centimeters. (Begin by measuring and recording the diameter of the cup's top in centimeters).

Cup diameter = _____ cm

Surface area of water = _____ cm²

Step 6: Compute the energy in calories that was collected in the cup ($Q = mc\Delta T$). Assume that the specific heat of the mixture is the same as the specific heat of the water. Show your work on separate paper, and record your result here.

Energy = _____ cal

Step 7: Compute the solar energy flux, the energy collected per square centimeter per minute.

Solar energy flux = _____ $\dfrac{cal}{cm^2\ min}$

Step 8: Compute how much solar energy reaches each square meter of Earth's surface per minute. Again, show your work on separate paper. (*Hint*: There are 10,000 cm² in 1 m².)

Solar energy flux = _____ $\dfrac{cal}{m^2\ min}$

Step 10: The distance between Earth and the Sun is 1.5×10^{11} m, or 1.5×10^{13} cm. The area of a sphere is $4\pi r^2$ so you can calculate the surface area of a "basketball" with radius 1.5×10^{13} cm. And you can divide this area by the surface area of the cup used in this experiment to find what fraction of the Sun's total solar output reached this cup! Then you can calculate the total solar output per minute from your work in Step 7. Do so.

Total solar output per minute = _____ cal

Summing Up

1. Scientists have measured the amount of solar energy flux just above our atmosphere to be 2 calories per square centimeter per minute (equivalently, 1.4 kW/m²). This energy flux is called the *solar constant*. Only 1.5 calories per square centimeter per minute reaches Earth's surface after passing through the atmosphere. How did your value compare?

    _____

    _____

2. What factors could affect the amount of sunlight reaching Earth's surface, decreasing the solar constant?

    _____

    _____

Name _____ Section _____ Date _____

CONCEPTUAL PHYSICAL SCIENCE

Driving Forces of Weather   **The Coriolis Effect**

Why Winds Wind Around Earth

Purpose
To investigate why Earth's rotation makes the global winds turn sideways

Apparatus
globe
poster board square (1/4 of a large sheet)
marking pen
baking soda
sieve or salt shaker
marble

Discussion
If Earth did not rotate, then air would simply flow from areas of high pressure to areas of low pressure. But Earth does rotate, or spin on its axis. How does this affect the flow of air—wind?

Procedure
Step 1: Look down at a globe from above, as though you were looking at the North Pole from outer space. Give the globe a gentle push so that it spins counterclockwise. Earth rotates in this direction—counterclockwise when viewed from over the North Pole

Step 2: You will use a piece of poster board to model Earth. Write "N" on one side of the board to represent the North Pole. Write "S" on the other side of the board to represent the South Pole.

Step 3: Give the poster board a spin counterclockwise. Practice this so that the board spins as smoothly as possible.

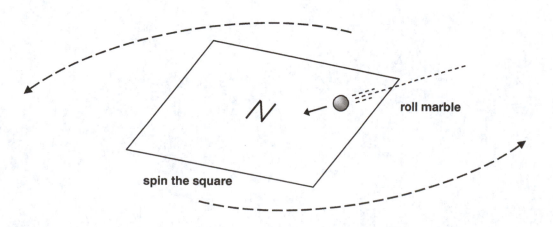

Figure 1.

Step 4: Sprinkle some baking soda on the North side of your poster board. Dust the board evenly and lightly.

Step 5: Wet the marble. Roll the marble across the poster board without rotating it.

    1.   Describe the path that the marble took in the powder. Is it straight or curved?

_____

_____

Step 6: Now, while a partner spins the poster board Earth model to simulate Earth's rotation, roll the wet marble across it.

    2.   Viewed from above, does the marble appear to follow a straight path or a curved path?

_____

_____

    3.   Is the path that the marble made in the powder straight or curved?

_____

_____

Step 7: Try rolling the marble from different directions across the spinning poster board. Look at the direction in which the marble moves.

    4.   Does the marble always turn in the same direction? Does it turn to the right or to the left (when viewed from above)?

_____

_____

Summing Up

1. Suppose you were the size of an ant, and you were riding on the poster board as it rotated. Would the marble appear to travel straight relative to you, or would it appear to turn?

    _____

    _____

2. If the poster board represents Earth in this activity, what does the marble represent?

    _____

    _____

3. As air molecules flow from the North Pole toward the equator, will they move in a straight path, or will they appear to turn (as seen by a viewer on the ground)? Why?

    _____

    _____

4. The apparent deviation of unattached objects traveling over Earth from the straight-line path is called the Coriolis Effect. How does the Coriolis Effect change the direction of global winds?

    _____

    _____

5. Air molecules don't _really_ turn as they move over Earth's rotating surface. Actually, they just appear to turn; the molecules are moving in a straight line. What physics law tells us that all moving objects move in straight lines if they are not acted upon by an outside force?

    _____

    _____

6. As viewed at Earth's surface, do ocean currents appear to turn as they move from the poles to the equator? If so, why?

    _____

    _____

WHY WINDS WIND AROUND EARTH

| **CONCEPTUAL PHYSICAL SCIENCE** | **Activity** |

Indoor Clouds

Purpose
To illustrate the formation of a cloud from the condensation of water droplets

Apparatus
large glass jar   measuring cup for water
small metal baking tray   tray of ice cubes

Discussion
Clouds are made up of millions of tiny water droplets and/or ice crystals. Cloud formation takes place as rising moist air expands and cools.

As the Sun warms Earth's surface, water is evaporated from the oceans, lakes, streams, and rivers. The process of evaporation changes the water molecules from a liquid phase to a vapor phase. Because evaporation is greater over warmer waters than over cooler waters, tropical locations have a higher water vapor content than polar locations. Water vapor content is dependent on temperature. For any given temperature, there is a limit to the amount of water vapor in the air. This limit is called the **dew point**. When this limit is reached, the air is saturated. A measure of the amount of water vapor in the air is called **humidity** (the mass of water per volume of air). The amount of water vapor in the air varies with geographic location and may change according to temperature.

Warm air can hold more water vapor than cooler air before becoming saturated. When the air temperature falls, the air becomes saturated—the amount of water vapor that the air can hold reaches its limit. As the air cools below the dew point, the water vapor molecules condense onto the nearest available surface. Condensation is the change from a vapor phase to a liquid phase. The condensation of water vapor molecules on small airborne particles produces cloud droplets, which in turn become clouds.

Most clouds form as air rises, expands, and cools. There are several reasons for the development of clouds. When the temperatures of certain areas of Earth's surface increase more readily than the temperatures of other areas, air may rise from thermal convection. As this warm air rises, it mixes with the cooler air above and eventually cools to its saturation point. The moisture from the warm air condenses to form a cloud. As the cloud grows, it shades the ground from the Sun. This cuts off the surface heating and upward thermal convection—the cloud dissipates. After the cloud is gone, the ground once again heats up to start another cycle of thermal convection.

Clouds also form as a result of topography. Air rises as it moves over mountains. As it rises, it cools. If the air is humid, clouds form. As the air moves down the other side of the mountain, it warms. This air is drier because most of the moisture has been removed to form the clouds on the other side. It is, therefore, more common to find cloud formation on the windward side of mountains than on the leeward side.

Clouds may form as a result of converging air. When cold air moves into a warm air mass, the warm air is forced upward. As it rises, it cools, and water vapor condenses to form clouds. Cold fronts are associated with extensive cloudiness and thunderstorms. When warm air moves into a cold air mass, the less dense warmer air rides up and over the colder, denser air. Warm fronts result in widespread cloudiness and light precipitation that may extend for thousands of square kilometers.

Procedure

Step 1: Fill a glass jar with about one inch of very hot water. Do not use boiling water—the glass may break.

Step 2: Place ice cubes in a metal baking tray, and set the tray on top of the jar. Make sure there is a good seal.

Step 3: Observe a "cloud." As the ice above cools the air inside the top of the jar, water vapor condenses into water droplets to form a "cloud."

Summing Up

1. How is your formation of clouds the same as the formation of clouds in the sky?

_____

_____

2. How is it different?

_____

_____

3. Why are coastal tropical areas more humid than desert areas?

_____

_____

4. What real-world observations make us believe that warm air rises?

_____

_____

5. How does topography contribute to the formation of desert areas? For example, what role does the mountain range we know as the Sierra Nevada play in the persistence of the Nevada desert?

_____

CONCEPTUAL PHYSICAL SCIENCE	**Activity**

The Solar System   **Orbital Geometry**

Ellipses

Purpose
To investigate the geometry of the ellipse

Apparatus
about 20 cm of string   2 thumbtacks
pencil and paper   a flat surface that will accept thumbtack punctures

---

Discussion

An ellipse is an oval-like closed curve defined as the locus of all points about a pair of foci (focal points), the sum of whose distances from those foci is a constant. Planets orbit the Sun in elliptical paths, with the Sun's center at one focus. The other focus is a point in space, typified by nothing in particular.

Elliptical trajectories are not confined to "outer space." Toss a rock, and the parabolic path it seems to trace is actually a small segment of an ellipse. If extended, its path would continue through Earth, swing about Earth's center, and return to its starting point. In this case, the far focus is Earth's center. The near focus is not typified by anything in particular. The ellipse is very stretched out, with its long axis considerably longer than its short axis. We say the ellipse is very *eccentric*. Toss the rock faster, and the ellipse is wider—less eccentric. The far focus is still Earth's center, and the near focus is nearer to Earth's center than before. Toss the rock at 8 km/s and both foci will coincide at Earth's center. The elliptical path is now a circle—a special case of an ellipse. Toss the rock faster, and it follows an ellipse external to Earth. Now, the near focus is the center of Earth, and the *far* focus is beyond—again, at no particular place. As speed increases and the ellipse becomes more eccentric again, the far focus is *outside* Earth's interior.

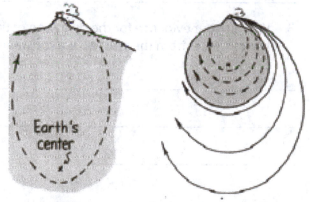

Constructing an ellipse with pencil, paper, string, and tacks is interesting. Let's do it!

Procedure

Step 1: Place a loop of string around two tacks or push pins, and pull the string taut with a pencil.

Step 2: Slide the pencil along the string, keeping it taut. (To avoid twisting of the string, it helps if you make the top half and the bottom half in two separate operations.)

Step 3: Label each focus of your ellipse (the location of the push pins).

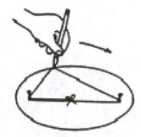

ELLIPSES

Step 4: Repeat, using different focus separation distances, and you will notice that the greater the distance between foci, the more eccentric the ellipse.

Step 5: Construct a circle by bringing both foci together. A circle is a special case of an ellipse.

Another special case of an ellipse is a straight line. Determine the positions of the foci that give you a straight line.

The shadow of a ball face-on is a circle. But when the shadow is not face-on, the shape is an ellipse. The photo below shows a ball illuminated with three light sources. Interestingly enough, the ball meets the table at the focus for all three ellipses.

Summing Up

1. Which of your drawings approximates Earth's orbit about the Sun?

_____

2. Which of your drawings approximates the orbit of Comet Halley about the Sun?

_____

3. What is your evidence for the definition of the ellipse: that for all points on the curve, the sum of the distances from the two foci is a constant?

_____

_____

_____

CONCEPTUAL PHYSICAL SCIENCE	Activity

The Solar System   **Pinhole Sun Projection**

Image of the Sun

Purpose
To produce an image of the Sun on your classroom floor

Apparatus
index card
sharp point (pencil, pen, or equivalent for poking a hole in the index card)
small knife (penknife, X-acto knife, or equivalent for cutting a hole in the index card)
meterstick
access to direct sunlight

Discussion

A pinhole camera is a light-tight box with a small pinhole-sized opening in one end and a viewing screen at the opposite end. Light coming from an object that passes through the pinhole forms an image of the object. Why? Because rays of light coming from the object cannot overlap and cause blurring. You can see in the sketch why the image is upside down.

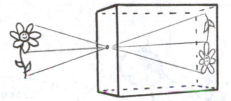

In this activity, your **pinhole camera** will simply be an index card with a hole punched in the middle by a pencil or pen. The object will be the Sun, bright enough to cast a clear image without a light-tight box.

Procedure

Step 1: Use the pencil or pen to poke a small hole in the center of the index card. The hole should be about one millimeter in diameter.

Step 2: Hold the card in bright sunlight about a meter or so above the floor. The card casts a shadow on the floor. In the middle of the card's shadow is a circular spot of light—an image of the Sun. If the Sun is low in the sky, the circular image is lengthened to an ellipse.

Step 3: Hold the card higher, and the solar image becomes bigger.

Step 4: Repeat the procedure, but instead of a round hole poked in the card, carefully cut a diamond shape.

Does this affect the shape of the image?

_____

_____

_____

Step 5: Measure the diameter of the solar image, and compare it to the distance between it and the pinhole. A convenient way to do this is to draw a circle 1 centimeter in diameter on a piece of paper. Place the paper on the floor where your solar image will fall. Then see how many centimeters high the pinhole needs to be for the solar image to fill the 1-cm circle.

How high is the pinhole above the paper? _____

Step 6: Determine and record how high the pinhole needs to be to produce a 2-cm-wide image.

_____

Step 7: Determine and record how high the pinhole needs to be to produce a 10-cm-wide image.

_____

Look at spots of light beneath sunlit trees. When the openings between leaves above are small compared with the distance to the ground below, images of the Sun are cast. What will be the shape of these images at the time of a partial solar eclipse?

CONCEPTUAL PHYSICAL SCIENCE	**Activity**

The Solar System   **The Size of the Sun**

Sunballs

Purpose
To estimate the diameter of the Sun

Apparatus
small piece of cardboard (such as an index card)   1 dime   meterstick

---

Discussion
Take notice of the round spots of light on the shady ground beneath trees. These are sunballs—images of the Sun. They are cast by openings between leaves in the trees that act as pinholes. The diameter of a sunball depends on its distance from the small opening that produces it. Large sunballs, several centimeters or so in diameter, are cast by openings that are relatively high above the ground, while small ones are produced by closer "pinholes." The interesting point is that the ratio of the diameter of the sunball to its distance from the pinhole is the same as the ratio of the Sun's diameter to its distance from the pinhole.

Because the Sun is approximately 150,000,000 km from the pinhole, careful measurement of this ratio tells us the diameter of the Sun. That's what this experiment is all about. Instead of finding sunballs under the canopy of trees, you will make your own easier-to-measure sunballs.

Procedure
Step 1: Poke a small hole in a piece of cardboard with a pen or sharp pencil. Hold the cardboard in the sunlight, and note the circular image that is cast on a convenient screen of any kind. This is an image of the Sun. Unless you are holding the card too close, note that the solar image size does not depend on the size of the hole in the cardboard (pinhole), but only on its distance from the pinhole to the screen. The greater the distance between the image and the cardboard, the larger the sunball.

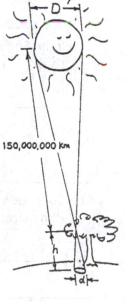

150,000,000 km

Step 2: Position the cardboard so the image exactly covers a dime, or something else that can be accurately measured. Carefully measure the distance to the small hole in the cardboard. Record your measurements as a ratio:

$$\frac{\text{Diameter of dime}}{\text{Distance from dime to pinhole}} = \underline{\hspace{3cm}}$$

Because this is the same ratio as the diameter of the Sun to its distance,

$$\frac{\text{Diameter of dime}}{\text{Distance from dime to pinhole}} = \frac{\text{diameter of Sun}}{\text{distance from Sun to pinhole}}$$

Which means that you can now calculate the diameter of the Sun!

Diameter of the Sun = _____

Summing Up

1. Will the sunball still be round if the pinhole is square-shaped? Triangular? (Experiment and see!)

_____

_____

_____

2. If the Sun is low in the sky so that the sunball is elliptical, should you measure the small or the long width of the ellipse for the sunball diameter in your calculation of the Sun's diameter? Why?

_____

_____

_____

3. If the Sun is partially eclipsed, what will be the shape of the sunball? (*Note*: The answer to this is shown in Figure 26.37 of your *Conceptual Physical Science* text!)

_____

_____

_____

4. Suppose you could fit 100 dimes, end to end, between your card with the pinhole and your dime-sized sunball. How many suns could fit between Earth and the Sun?

_____

_____

_____

CONCEPTUAL PHYSICAL SCIENCE	Activity

The Solar System   **A Planetary Puzzle**

Tracking Mars

Purpose

To plot the orbit of Mars employing the technique of Johannes Kepler and the data obtained by Tycho Brahe four centuries ago

Apparatus

four sheets of plain grid graph paper   compass   protractor
ruler   sharp pencil

Discussion

In the early 1500s, the Polish astronomer Nicolaus Copernicus used observations and geometry to determine the orbital radius and orbital period for the planets known at the time. Copernicus based his work on the revolutionary assumption that the planets moved around the Sun. In the late 1500s, before telescopes were invented, the Danish astronomer Tycho Brahe made 20 years of extensive and accurate measurements of planets and bright stars. Near the end of Tycho's career, he

hired a young German mathematician, Johannes Kepler, who was assigned the task of plotting the orbit of Mars using Tycho Brahe's data.

Kepler started by drawing a circle to represent Earth's orbit (not bad, considering that Earth's orbit only approximates a circle, which he did not know at the time). Since Mars takes 687 Earth days to orbit the Sun once, Kepler paid attention to observations that were exactly 687 days apart. In this way, Mars would be in the same place while Earth would be in a different location. Two angular readings of the location of Mars, from the same location on Earth 687 days apart, were all that was needed—where the two lines crossed was a point on the orbit of Mars. Plotting many such points did not trace out a circle, as Kepler had expected—rather, the path was an ellipse. Kepler was the first to discover that if the planets orbited the Sun, they did so in elliptical rather than circular paths. He then went on to plot a better orbit for Earth based on observations of the Sun, and further refined the plotted orbit of Mars.

In this activity, you will duplicate the work of Kepler, simplifying somewhat by assuming a circular orbit for Earth—it turns out the difference is minor, and the elliptical path of Mars is evident. From Brahe's extensive tables, we will use only the data shown in Table A.

Procedure

Step 1: You will want a sheet of graph paper with about a 14 × 14-inch working area. If need be, tape two legal-size sheets of borderless graph paper together, or four regular 8 1/2" × 11" sheets.

Step 2: Make a dot at the center of your paper to represent the Sun. Place a compass there, and draw a 10-cm-radius circle to represent the orbit of Earth around the Sun. Draw a light line from the center to the right of the paper, and mark the intersection with Earth's orbit 0°. This is the position of Earth on September 21. All your plotting will be counterclockwise around the circle from this reference point.

Table A. Brahe's data are grouped in 14 pairs of Mars sightings. For the first 9 pairs, the first line of data is for Mars in opposition—when Sun, Earth, and Mars were on the same line—when Mars was 90° to the Earth horizon, directly overhead at midnight. The second line of data are positions measured 687 days later, when Mars was again in the same place in its orbit, and Earth in a different place, where a different angle was then measured. All angles given in the table read from a 0° reference line—the line from the Sun to Earth at the autumnal equinox, September 21. Mars at points 10–14 are non-opposition sightings. The first line of Point 10, for example, shows that when Earth was at 277°, Mars was seen not directly overhead, but at 208.5° with respect to the 0° reference line. Then, 687 days later, Earth was at 235°, and Mars was seen at 272.5°. The data are neatly arranged for plotting—something that took Kepler years to do.

Step 3: Locate the first point in Mars's orbit, Point 1, from Table A. Do this by first marking with a protractor the position of Earth along the circle for the date November 28, 1580. This is 66.5° above the 0° reference line. Draw a dot to show Earth's position at this time.

Mars Orbit Point	Date			Earth Position (Ecliptic)	Mars Position (Ecliptic)
	Mo	Day	Year		
1	11	28	1580	66.5	66.5
	10	16	1582	22.5	107
2	1	7	1583	107	107
	11	24	1584	62.5	144
3	2	10	1585	141.5	141.5
	12	29	1586	97.5	177
4	3	16	1587	175.5	175.5
	1	31	1589	132	212
5	4	24	1589	214.5	214.5
	3	12	1591	171.5	253.5
6	6	18	1591	266.5	266.5
	5	5	1593	225	311
7	9	5	1593	342.5	342.5
	7	24	1595	300.5	29.5
8	11	10	1595	47.5	47.5
	9	27	1597	4	90
9	12	24	1597	92.5	92.5
	11	11	1599	48	130.5
10	6	29	1589	277	208.5
	5	16	1591	235	272.5
11	8	1	1591	308	260.5
	6	18	1593	266.5	335
12	9	9	1591	345.5	273
	7	27	1593	304	347.5
13	10	3	1593	9.5	337.5
	8	20	1595	327	44.5
14	11	23	1593	60.5	350.5
	10	10	1595	17	56

Step 4: Mars at this time was in *opposition*—opposite to the Sun in the sky. A line from the Sun to Earth at this time extends radially outward to Mars. Draw a line from the center of your circle (the Sun) to Earth's position at this time, and beyond Earth through Mars. Where is Mars along this line? You will need another sighting of Mars 687 days later, when Mars is at the same place and Earth is at another.

Step 5: If you were to add 687 days to November 28, 1580, you would get October 16, 1582. At that date, Mars was measured to be lower in the sky—actually 107.0° with respect to the 0° reference line of September 21. With respect to the reference line, use a protractor and ruler and draw a line at 107.0° as shown in Figure 1. Where your two lines intersect is a point along the orbit of Mars.

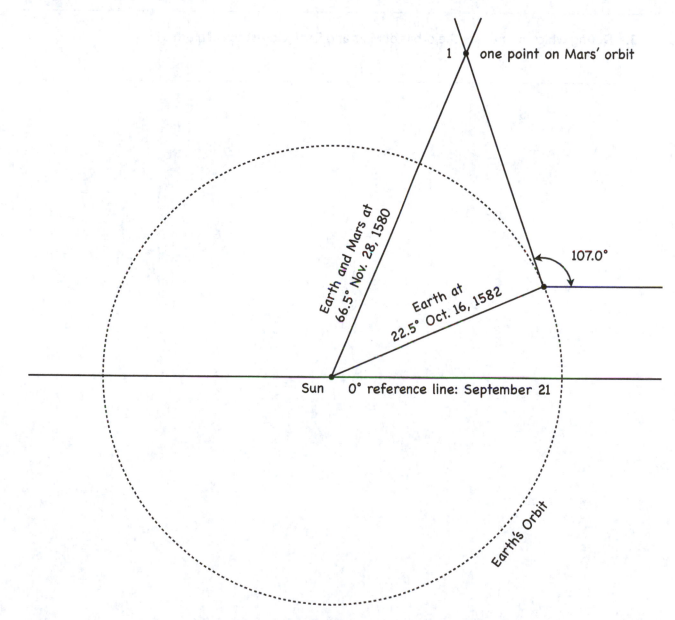

Figure 1. Your plot should look like this for finding Mars' Orbit Point 1; data are in the first pair of lines of Table A. Angles from Earth to Mars are given with respect to the 0° reference line of September 21.

Step 6: With care, plot the 13 other intersections that represent points along the orbit of Mars, using the data in Table A.

Step 7: Connect your points, either very carefully by freehand or with a French curve.

Bravo—you have plotted the orbit of Mars using Kepler's method from four centuries ago!

TRACKING MARS

Summing Up

1. Place the point of your compass on the position of the Sun, and place the pencil on the first point of the plotted orbit of Mars. Draw a circular orbit with that radius. Does that circle match the orbit of Mars? If not, describe the differences.

_____

2. Does your plot agree with Kepler's finding that the orbit is an ellipse?

_____

3. During which month are the orbits of Mars and Earth closest to each other?

_____

Name _____ Section _____ Date _____

CONCEPTUAL PHYSICAL SCIENCE | Experiment

Stars and Galaxies   **Observing the Sky**

Reckoning Latitude

Purpose

To build and use the instruments necessary for determining how far above the equator you are on this planet

Apparatus

protractor
pencil with unused eraser
pushpin
drinking straw
plumb line and bob
tape

Discussion

Latitude is a measure of your angular distance north or south of the equator (Figure 1). To determine your latitude in the Northern Hemisphere, it is convenient to use Polaris, the North Star, as a guide. At the North Pole, you would find that the North Star was directly above you—90° above the horizon. This is also your angular distance from the equator (Figure 2a). If you were to travel closer to the equator, the North Star would recede behind you to some lower angle above the horizon, which would always match your angular distance from the equator (Figure 2b). For example, if you measured the North Star to be 45° above the horizon, you would be at an angular distance of 45° north of the equator. At the equator, you would find the North Star to be 0° above the horizon (Figure 2c). (It turns out that atmospheric refraction at this grazing angle complicates viewing.)

Figure 1. Latitudes are parallel to the equator.

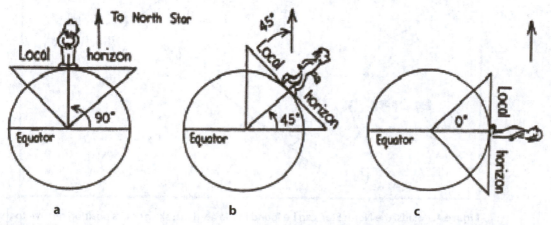

Figure 2. The angle at which the North Star appears above the horizon is also your angular distance from the equator.

RECKONING LATITUDE

Procedure

An astrolabe is a simple instrument for measuring angles above the horizon (altitude). To construct an astrolabe, tape a straw to the straight edge of a protractor, as shown in Figure 3. Place the eraser end of a pencil against the hole in the protractor, and insert the pushpin through the other side of the hole into the eraser—the pencil serves as a convenient handle. Next, tie a plumb bob to the pushpin (a string attached to a weight).

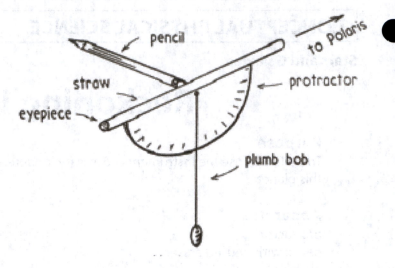

To measure the altitude of any object viewed through the straw, have a partner read the angle that the plumb bob line makes with the protractor. Practice using your astrolabe by measuring the angle of various objects around you. Work in pairs or small groups so that one person can read the angle while the other is sighting. Take turns looking through the astrolabe to be sure of your readings. You may want to make several measurements and find the average. The angle that you read should be between 0° and 90°. Be sure that this angle is not the number of degrees from the zenith (straight up). You can convert such readings into degrees from the horizon (straight out) by subtracting the number of degrees from 90°. Challenge yourself or others to estimate measurements without using the astrolabe. How accurate are these "naked-eye" estimates?

Measuring the Altitude of Polaris

The North Star is fairly easy to locate because of its position relative to the Big and Little Dippers, two well-known constellations that stand out in the northern sky. To find the North Star from the Big Dipper, draw an imaginary line between the last two stars of the Big Dipper's pan. Follow this line away from the top of the Big Dipper, and the first bright star you cross is the North Star (Figure 4). Note that the North Star is the first star of the Little Dipper's handle.

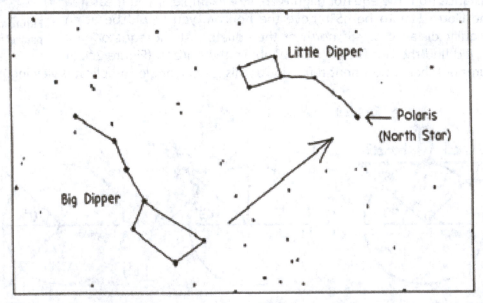

Figure 4. Polaris, the North Star, can be found in the northern sky from its position relative to the Big and Little Dippers.

To find your latitude, you need simply measure the altitude of the North Star above the horizon. It happens that the North Star is not *exactly* above the North Celestial Pole, so for a more accurate determination of your latitude, a minor but significant correction may be necessary. The North Star lies about 3/4 of a degree away from the North Celestial Pole in the direction of Cassiopeia, a "W"-shaped constellation (Figure 5). Hence the altitude of the North Star may not correspond exactly to your latitude. If you see Cassiopeia much "above" the North Star (higher up from the horizon), then subtract 3/4 of a degree from your measured altitude of Polaris to obtain your latitude. If, on the other hand, Cassiopeia is "beneath" the North Star, add 3/4 of a degree. If Cassiopeia is anywhere to the left or right of the North Star, then the correction is not necessary.

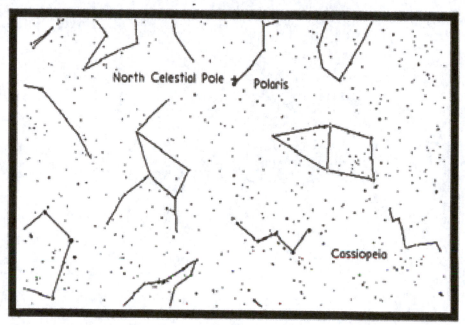

Figure 5. Polaris is offset from the pole toward Cassiopeia.

Summing Up

1. What was your measured altitude of Polaris?

_____

2. According to this measurement, at what latitude are you located?

_____

3. By how many degrees does your measured latitude differ from the accepted latitude given by your instructor?

_____

4. One degree of latitude is equivalent to 107 km (66 miles). By how many kilometers does your determined latitude differ from the accepted value in kilometers?

    (Your determined latitude – accepted latitude) × 107 = _____

5. How could the astrolabe be used to provide evidence that the world is round?

_____

Appendix A: Physical Quantities and Units of Measure

These appendices are not definitive, complete references on the subjects they address. Rather, they are introductions to these topics, intended to serve the purposes of this manual. They are simplified and leave much to be discovered in later coursework and practice.

Quantities and Units Are Two Different Things

One challenge when working in physical science is distinguishing between physical quantities and units of measure. When first learning about these things, it's easy to confuse "force" with "newtons" or "power" with "watts." Physical quantities and units of measure *are* related. But just like you and your cousin, they are distinctly different from one another.

Physical quantities are any of the characteristics of objects or events in nature that can be measured.

This lab manual, for example, is an object with several quantifiable characteristics. Its dimensions can be measured; such measurements involve the physical quantity of *length*. Other physical quantities can also be measured: The manual's *mass* can be measured, its *weight* can be measured, its *volume* can be measured, its *temperature* can be measured, and so on. Length, mass, weight, volume, and temperature are *physical quantities*.

If physical quantities are to be quantified, they must be measured in finite units. **Units of measure** are the finite standards of measurement, such as meters, kilograms, newtons, cubic meters, and kelvins.

Different physical quantities correspond to different units of measure. The *quantity* of length can be measured in *units* of meters, feet, miles, or light-years, for example, but not in units of kilograms, seconds, watts, or ounces. A quantity of time can be measured in seconds, but not in meters or pounds.

As you work in physical science, it is important that you distinguish between physical quantities and units of measure and use correct terminology in appropriate contexts. We may say, "The force acting on the object was 5 newtons," but we do not say, "the newtons acting on the object was 5." We say, "The volume of liquid was 50 cubic centimeters," but not "The cubic centimeters of the liquid was 50." A moving object has a speed; it does not have a miles-per-hour. And so on.

Below is a table showing a few physical quantities and some corresponding units of measure.

Physical Quantity (*symbol**)	Unit of Measure (abbreviation*)
distance (*d*) or length (*L*)	meter (m), foot (ft), light-year (ly), inch (in), angstrom (Å)
time (*t*)	second (s), hour (h), day (d), year (y)
mass (*m*)	kilogram (kg), slug (slug)
weight (*W*)	newton (N), pound (lb)

*Symbols for quantities are *italicized,* but abbreviations for units are not: *m* is used for mass, while m is used for meters.

Formulas in science are written in terms of physical quantities. Formulas are never written in terms of units. The formula for speed is $v = d/t$ and is never expressed as $v = m/s$ or $v = mi/h$.

If you can't distinguish between quantities and units, you could run into confusion when dealing with formulas.

Consider the formula for density: $D = m/V$.

If you're on top of the distinction between quantities and units, then you have no problem: D is density, m is mass, and V is volume. If you can't distinguish between units and quantities, you might decide that m is meters and thus density is length/volume instead of mass/volume.

The International System of Units: Système International d'Unités (SI)

Scientific measurements are made in units of measure that are commonly used worldwide. This family of units is often referred to as the metric system. The more accurate term is *SI units*, where SI is an abbreviation for the French term for "International System." SI units are preferred over Imperial or US Customary units, such as feet and pounds. Imperial units originated in the United Kingdom, but they remain popular in few places outside the United States. The world at large, including the UK, has moved to SI units.

Base Quantities and SI Base Units

Base units correspond to the most fundamental physical quantities. Each base quantity is independent of all the other base quantities, but the units are interdependent on one another.

Quantity	Symbol	SI Unit	Abbrev.	Notes
distance	d	meter	m	Originally one ten-millionth of one quarter of Earth's circumference, now how far light travels in 1/299,792,458 of a second.
mass	m	kilogram	kg	Originally the mass of a liter of water, now the mass of an iridium-platinum prototype.
time	t	second	s	Originally 1/86,400 of a day, now the interval required for a cesium-133 atom to vibrate 9,192,631,770 times.
temperature	T	kelvin	K	The fraction 1/273.16 of the thermodynamic temperature of the triple (freezing) point of water.
electric current	I	ampere	A	The current needed to produce 2×10^{-7} N/m of magnetic force between two long, straight, current-carrying wires 1 meter from one another.
quantity of substance	N	mole	mol	The number of atoms in 12 grams of carbon-12. The 2011 value is $6.02214078 \times 10^{23} \pm 0.00000018 \times 10^{23}$.
luminous intensity	L	candela	cd	This unit is not typically used in introductory science courses.

Some Derived Quantities and the Corresponding Derived Units

Other quantities are also needed to describe and measure objects and events. The following quantities are derived from the base quantities through multiplication or division.

Quantity	Symbol	Unit	Abbrev.	Alternate(s)
area	A	square meter	m^2	
volume	V	cubic meter	m^3	
frequency	f	hertz	Hz	1/s
speed	v	meter per second	m/s	
acceleration	a	meter per second squared	m/s^2	
force	F	newton	N	kg·m/s
momentum	p	kilogram-meter per second	kg·m/s	N·s
work, energy, heat	W, E, Q	joule	J	N·m, $kg·m^2/s^2$
power	P	watt	W	J/s, $kg·m^2/s^3$
charge	q	coulomb	C	A·s
potential	V	volt	V	J/C
resistance	R	ohm	Ω	V/A

Appendix B: SI Prefixes and Conversion Factors

SI Prefixes

In the metric (SI—International System) system of units, the power-of-10 notation can be replaced by prefixes that denote a certain power of 10. For example, kilo denotes 10^3 and micro denotes 10^{-6}. Each prefix also has an abbreviation. The abbreviation for kilo is k, and the abbreviation for micro is μ. Thus 1000 g (one thousand grams) is 10^3 g or 1 kg (one kilogram), and 0.000001 m (one millionth of a meter) is 10^{-6} m or 1 μm (one micrometer). There is an SI prefix for every third power of 10. Here are a few of them.

Prefix	Pronunciation	Abbrev.	Value	Prefix	Pronunciation	Abbrev.	Value
pico	PEE koe	p	10^{-12}	**kilo**	KEE loe	k	10^3
nano	NAH noe	n	10^{-9}	**mega**	MEH guh	M	10^6
micro	MY kroe	μ	10^{-6}	**giga**	JEE guh*	G	10^9
milli	MIH lee	m	10^{-3}	**tera**	TARE uh	T	10^{12}

*Some computer experts say "GIH guh," but they also tend to spell disc as "disk," McIntosh as "Macintosh," and Googol as "Google." Proficiency with computers and proficiency with language rely on somewhat different skill sets.

Some Useful Conversion Factors

This list is by no means comprehensive. It includes several factors that may be useful in one or more of the lab activities.

centimeters and meters	1 cm = 0.01 m	and	1 m = 100 cm
inches and meters	1 in = 0.025 4 m	and	1 m = 39.37 in
feet and meters	1 ft = 0.308 4 m	and	1 m = 3.281 ft
yards and meters	1 yd = 0.914 4 m	and	1 m = 1.094 yd
grams and kilograms	1 g = 0.001 kg	and	1 kg = 1000 g
miles per hour and meters per second	1 mph = 0.447 m/s	and	1 m/s = 2.24 mph
cubic centimeters and cubic meters	1 cm^3 = 0.000 001 m^3	and	1 m^3 = 1 000 000 cm^3

Appendix C: Accuracy, Precision, and Error

Accuracy and Precision

Suppose the mass of a gold coin were measured using the best scale available. The resulting measurement would be both accurate and precise. Suppose that value were 27.96458 grams. If measured with the best scale available, that measurement would be considered the accepted value.

A second scale, known to be of lesser quality, gives a value of 28 grams. This measurement would best be described as being accurate but not precise. It is close to the accepted value, but it has only two significant figures. **Being close to the accepted value makes it accurate.**

A third scale, again of lesser quality than the original, gives a value of 35.2156 grams. This measurement would best be described as precise but not accurate. It has six significant figures, but it isn't very close to the accepted value. **Having many significant figures makes a value precise.**

A fourth scale, of truly inferior quality, gives a value of 15 grams. This measurement is neither accurate nor precise.

A measurement is considered accurate if it is close to the accepted value. A measurement is considered precise if it includes many significant figures (measures a value across many orders of magnitude).

Random and Systematic Error

One cannot make a measurement without error. There is error in every measurement. Error can be minimized, but not eliminated. In laboratory measurements, there are generally two kinds of error: random and systematic. The meanings of and differences between random error and systematic error can be illustrated using an archery analogy.

During target practice, an archer launches 10 arrows at a target. The archer hopes to get all 10 arrows to the center of the target. Consider the results from four archers.

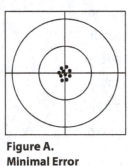
Figure A. Minimal Error

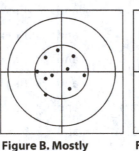
Figure B. Mostly Random Error

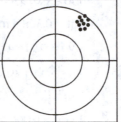
Figure C. Mostly Systematic Error

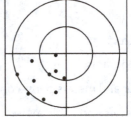
Figure D. Random and Systematic Error

Archer A's shots show very little error. They center on the "bull's eye" **and** form a tight cluster.

Archer B's shots suffer from random error. They center on the "bull's eye" but are scattered: some high, some low, some left, and some right. The error seems to be **different** with each shot.

Archer C's shots suffer from systematic error. Though the cluster is tight, they do not center on the "bull's eye." Each shot was high and to the right. The error seems to be **the same** with each shot.

Archer D's shots suffer from both random and systematic error. They are scattered **and** do not center on the "bull's eye."

If called upon to describe sources of error in an activity or experiment, be careful about citing **human error.** Students sometimes (understandably) confuse human error with mistakes. Error cannot be avoided, but mistakes cannot be accepted. When a mistake is made during a procedure, simply discard the resulting observations and repeat the step in careful accordance with the instructions.

Appendix D: Uncertainty and Significant Figures

Counting Numbers and Exact Values

Some numbers are exact. The number of eggs in a dozen is 12. There is no uncertainty; the value of 12 is exact. No measurement technology or skill will change the value. Many people have 10 fingers and 10 toes. There are 52 cards in a standard deck. A value need not be a whole number to be exact. The price of a gallon of gasoline is an exact value, even though it is often given with a decimal value, such as 2.659 USD (US dollars). Exact values are often stated, imposed, counted, or given. They are not measurements.

Uncertainty and Precision in Measurements

All measurements include some degree of uncertainty. This section is devoted to the rules used at the introductory level to respect the uncertainty in measurement.

You might use a meterstick—marked in millimeters—to measure the width of a table and find it to be 76.14 cm. If your meterstick were marked only in centimeters, you might find the width to be 76.1 cm. If you used a better device, you might find the width to be 76.138 235 cm. The 76.1-cm measurement is the least precise and has the greatest uncertainty. The 76.138 235-cm measurement is the most precise and has the least uncertainty.

The precision of a measurement is indicated by the number of significant figures it includes.

76.1 cm	three significant figures
76.14 cm	four significant figures
76.138 235 cm	eight significant figures

Some Rules of Significant Figures

1. In numbers that do not contain zeros, all the digits are significant.

EXAMPLES

4.132 7	five significant figures
5.14	three significant figures
369	three significant figures

2. All zeros between significant digits are significant.

EXAMPLES

8.052	four significant figures
7059	four significant figures
306	three significant figures

3. Zeros to the left of the first nonzero digit serve only to fix the position of the decimal point and are not significant.

EXAMPLES

0.006 8	two significant figures
0.042 7	three significant figures
0.000 350 6	four significant figures

4. In a number with digits to the right of the decimal point, zeros to the right of the last nonzero digit are significant.

EXAMPLES

53	two significant figures
53.0	three significant figures
53.00	four significant figures
0.002 00	three significant figures
0.700 50	five significant figures

5. In a number that has no decimal point and that ends in one or more zeros (such as 3600), the zeros that end the number may or may not be significant.

The number is ambiguous in terms of significant figures. Further information about how the number was obtained is needed before the number of significant figures can be specified. If it is a measured number, the zeros are not significant. If the number is a defined or counted number, all the digits are significant.

Confusion is avoided when numbers are expressed in scientific notation. All digits are taken to be significant when a number is expressed this way.

EXAMPLES

4.6×10^{-5}	two significant figures
4.60×10^{-5}	three significant figures
4.600×10^{-5}	four significant figures
2×10^{-5}	one significant figure
3.0×10^{-5}	two significant figures
4.00×10^{-5}	three significant figures

Rounding

Calculators often display eight or more digits. How do you round such a display to, say, three significant figures? Three rules govern the process of deleting unwanted (insignificant) digits from a calculator number.

1. If the first digit to the right of the last significant figure is less than 5, that digit and all the digits that follow it are simply dropped.

EXAMPLE

51.234 rounded to three significant figures becomes 51.2.

2. If the first digit to be dropped is a digit greater than 5, or if it is a 5 followed by a digit other than zero, the excess digits are dropped and the last retained digit is increased in value by one unit.

EXAMPLE

51.35, 51.359, and 51.359 8 rounded to three significant figures all become 51.4.

3. If the first digit to be dropped is a 5 not followed by any other digit, or if it is a 5 followed only by zeros, an odd-even rule is applied.

That is, if the last retained digit is even, its value is not changed, and the 5 and any zeros that follow are dropped. But if the last digit is odd, its value is increased by one. The intention of this odd-even rule is to average the effects of rounding off.

EXAMPLES

74.250 0 to three significant figures becomes 74.2.

89.350 0 to three significant figures becomes 89.4.

Appendix E: Percent Error and Percent Difference

Percent Error

There are times when you want to compare a value found through experiment to an accepted (or standard) value. The absolute difference between the two isn't always useful. For example, an experimental value of gravitational acceleration (9.8 m/s^2) that is off by 3 m/s^2 would be unimpressive. But an experimental value for the speed of light (299,792,458 m/s) that is off by 3 m/s would be very impressive indeed!

A better measure of error is to compare the error to the accepted value. One way to do this is to calculate percent error. For a given experimental value and accepted value, the percent error is

$$\% \text{ Error} = \frac{\text{measured} - \text{accepted}}{\text{accepted}} \times 100$$

Some prefer to express percent error as a positive value only. If you wish t o do so, simply take the absolute value of the result from the equation above. Ask your instructor for guidance.

Percent Difference

There are occasions when you want to compare two measurements. Since both values are measured, neither can act as the accepted standard. To calculate the percent difference, divide the difference between the measured values by their *average*.

The percent difference between two values, a and b, can be found by using the following equation.

$$\% \text{ Difference} = \frac{|a-b|}{(a+b)/2} \times 100 \qquad \text{This simplifies to} \quad \% \text{ Difference} = \left|\frac{a-b}{a+b}\right| \times 200.$$

Percent difference is expressed as a positive value.

Appendix F: Graphing

PREPARE THE GRAPH SPACE

1. Use as much of the available graphing space as you can. Leave room for labeling the scale, quantity, and units of each axis.

2. Label the scale of each axis. Label the origin as 0 for both axes. Select a scale so that the range of data fills the graph. Make sure your scales are manageable. Perhaps you might let four squares represent one whole unit of measurement (such as 1 meter or 1 volt). You probably won't want to let seven squares represent one unit. But again, fill as much of the page as is practical.

3. Label the quantity and units of each axis. Identify the quantity and then abbreviate the units in parentheses: "Time (s)" or "x (cm)," for example.

4. Title the graph. You may choose to give the graph a descriptive title, such as "Motion of a Toy Car." But you should always include a title bearing the names of the quantities, such as "Position vs. Time." In such a title, the first quantity listed is the one represented on the vertical axis.

PLOT THE DATA

1. Make a small but visible mark at each data point. If the point (0, 0) is a valid data point, be sure to mark it, too.

2. Resist the urge to connect the dots.

3. When you are finished, you will have what is called a scatter graph.

DRAW A LINE OF BEST FIT FOR A LINEAR PLOT (IF ASKED TO DO SO)

1. A line of best fit is not found by connecting the data point dots. Do **not** connect the dots!

2. If plotting a linear "best-fit" line, make a single straight line that comes as close as possible to all the data points. The line may not pass through any of the actual data points (depending on the amount of scatter in the data). It need not pass through the origin or the final data point either.

3. If multiple plots are to be graphed on a single set of axes, be sure to label each best-fit line to distinguish it from the others.

USE THE LINE OF BEST FIT

1. Once the line of best fit has been drawn, use that line to calculate the slope.

2. Disregard the slope between any two particular data points. The line of best fit is the only line to use for data analysis. It is as if the original data no longer exist, and only the line remains.

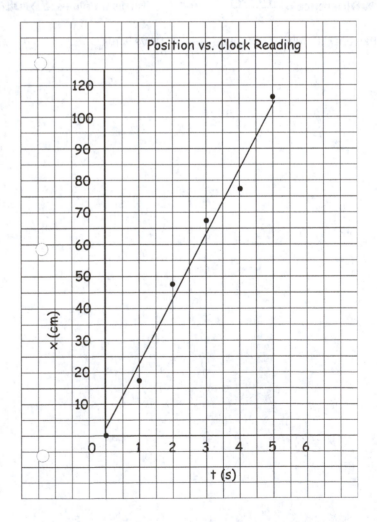